Access to Mobile Services

ADVANCES IN DATABASE SYSTEMS
Volume 38

Series Editors

Ahmed K. Elmagarmid
Purdue University
West Lafayette, IN 47907

Amit P. Sheth
Wright State University
Dayton, Ohio 45435

For other titles published in this series, go to
www.springer.com/series/5573

Access to Mobile Services

by

Xu Yang
Spirent Communications
USA

Athman C. Bouguettaya
CSIRO ICT, Australian National University
Australia

 Springer

Xu Yang
Spirent Communications
20324 Seneca Meadows Parkway
Germantown, MD 20876, USA
yangxu0@gmail.com

Athman Bouguettaya
CSIRO ICT Center
Computer Science and Information
  Technology Bldg.
Australian National University
North Road, Acton, ACT 2601, Australia

ISBN: 978-1-4419-4699-7   e-ISBN: 978-0-387-88755-5
DOI: 10.1007/978-0-387-88755-5

Printed on acid-free paper

springer.com

Foreword

Mobile technology has experienced an unparalleled growth in recent years, due to the development of new handheld devices and improved wide-area cellular data coverage and bandwidth and the seamless integration of wireless data access into PDAs that offer improved connectivity to the Internet. The recent advances in wireless technologies coupled with more powerful handhelds and cell phones as well as the convergence of voice, data, content, and mobile services have given birth to the mobile computing paradigm which has infiltrated all aspects of our lives. For instance, mobile services providers are deploying next-generation 3G and WiMax broadband technologies to deliver value-based content offerings, such as location-based advertising and mobile commerce to entice customers to use their services. Even traditional content providers, such as AOL and Yahoo, and consumer retailers are getting partnering with multi-vendor network operators who provide the underlying network infrastructure to drive their own branded communications services.

Advanced wireless applications are not standalone. They are part of a complex structure that spans wireless devices, wireless networks, the Internet, and back-end systems that typically reside on enterprise platforms. Creating networked wireless applications requires a broad set of skills, and knowledge of client technologies like the Mobile Information Device Profile (MIDP), a fundamental part of Java 2 Platform, Micro Edition (J2ME), of server technologies and of the mechanisms they use to communicate with each other. Enterprises can now attempt large scale deployments of field service, route management, and field sales projects, thus automating and tightly integrating a field-based workforce into a companys stationary IT infrastructure.

Recently, advances in wireless technologies and more powerful handhelds and cell phones have given birth to the Mobile Computing paradigm. Mobile Computing represents a fundamentally new paradigm in enterprise computing. It enables operating a job and role specific application loaded on a handheld or tablet device that passes only relevant data between a field worker and the relevant back-end enterprise systems regardless of connectivity availability. This comes in contrast to conventional distributed enterprise applications that can be accessed via a machine in a fixed network connection from a remote location. The overall goal is to provide users with universal and immediate access to information, no matter whether it is available on a fixed or wireless network, and to transparently support them in their tasks.

The Service Oriented Architecture approach is rapidly becoming the accepted standard for the future of Mobile Computing. With Web services being widely deployed as the Service-Oriented Architecture of choice for internal processes in organizations, there is also an emerging demand for using Web Services enabling mobile working, business-to-business use cases, and business-to-consumer use cases. Web services technology provides a way of connecting disparate environments, bridging mobile device applications, mobile network servers and application servers. This is achieved by passing standardized messages between services. As such services are available via standardized, open interfaces, mobile network operators will be able to offer services used both by applications on the phone, and other networked service and content providers. Mobile services are a breed of Web services that are accessible by mobile clients through wireless networks. Mobile services introduce a kind of anytime and anywhere access to services.

Mobile services introduce a variety of serious requirements for advanced applications when compared with their wired counterparts. These include the need for lightweight applications and non-interactive information-oriented services that usually retrieve real-time information that is usually accessed by large numbers of users. All these factors place serious demands for a relatively sophisticated mobile infrastructure that supports easy access, discovery, composition and flexible execution of mobile services for wireless broadcast environments. This is precisely the topic of this book.

This book is a thorough and detailed study of broadcast-based mobile service architectures and their application to modern mobile-based applications. The book considers broadcast to be the main method of delivering mobile-services to users and takes a close look at the new challenges for broadcast services, such as access patterns, service-data dependencies and access semantics and focuses on efficient access methods for broadcast based mobile-services that address these challenges. To achieve its objectives the book conducts an extensive study of several existing access techniques for information-oriented wireless networks and proposes novel data access methods that improve conventional access methods. It also develops a testbed for evaluating these data access methods, presents analytical cost models for each access method and conducts extensive experiments for comparing the existing and the proposed access methods. The book then presents novel wireless broadcast infrastructure that supports discovery and composition of mobile-services and define access semantics for this infrastructure and illustrates how to leverage these semantics to achieve best possible access efficiency. Subsequently, the book presents a practical study of the proposed infrastructure, access methods, and channel organizations and a testbed implementation for simulating accessing composite services in a broadcast-based environment. Finally, the book concludes by stating open research problems and future directions.

The book covers an impressive number of topics with great clarity and accuracy. The coverage is clear, logical and highlights key points well. It is good to see that

all difficult topics are explained in a lucid manner and all include extensive bibli-
ographies to help the interested reader find out more about these topics, particularly
where detailed explanations have been omitted for reasons of space. I commend the
authors on the breadth and scope of their work.

In summary, I highly recommend this book to anyone who wants to know more
about mobile computing and service. It is an invaluable read for advanced students,
researchers, and IT professionals. It is well thought out and eminently readable!

Tilburg, The Netherlands, November 2008   *Michael P. Papazoglou*

Foreword

Many recent technological and social trends have profoundly changed the way we live many aspects of our lives. For example, in a seemingly distant past, we used to plan. Plan evenings with friends, plan vacations, and plan a long software development process. Planning is now out of fashion in many domains. When we meet our friends for the evening, we do not always agree beforehand for where exactly to meet, and we do not look up the route to the location we need to go. We have our mobile phones to synchronize as we go on exactly where and when to meet, we have our GPS navigator that guides us, and we can book the restaurant (and even order food) on the fly with our smartphone. In software, the traditional waterfall software development processes are in many cases being replaced by agile methods, which can be (unfairly) summarized as do first, plan later. A similar trend is going on with respect to the availability of information. Waiting is no longer an option. Delays are no longer accepted. Information is available any time, anywhere. Today I can know where a flight is in the world, at what height it is and how fast is going. I can know the real-time quote of any stock in almost any stock exchange in the world. I can even know what my friends are up to, wherever they live. Technology made all this possible. And what is significant about it is that it affected what we come to expect in every context. Real-time anywhere is now the common expectation. Borrowing a famous sentence from an ad campaign, life is now. Not later. The straightforward conclusion is that mobile services are going to witness an incredible growth over the next few years. This growth is already happening, fueled by the needs I described above but also by the rapid diffusion of mobile phones and of smartphones, even in relatively poor countries. Even if we do not consider mobile phones, traveling with laptop is the norm, and between wireless and 3G devices integrated with the laptop, we have the potential of being connected all the time. Both the benefits to society and the revenue potential for businesses are enormous.

. This timely book tackles critical issues that may facilitate the adoption of and access to mobile services, focusing in particular on broadcast wireless networks. First, Yang and Bouguettaya discuss data access in these networks, by tackling three main aspects: i) provide an analytical model to evaluate various access methods, ii) offer a testbed where the methods can be experimented, and iii) provide e novel technique that is superior to what is available today and that adapt to the characteristics of the environment. All these are important contributions in their own right. Then, Yang and Bouguettaya move to consider novel problems in the context of mobile services: composition and semantics. Specifically, the authors show how to provide, within a broadcast wireless environment, a system that supports the discovery of mobile services as well as their composition. An interesting aspect here relates to how the information of a broadcast feed should be structured so that users can easily discover and compose services. Semantic aspects, such as semantic access to composite web services to achieve the best possible service quality and performance are also considered and they represent a quite original aspect of this book. I think you will find this book interesting and pleasant to read, and I hope you enjoy it.

November 2008   *Fabio Casati*

Preface

The emerging field of service computing is undeniably one of the few recent innovations in computing that are poised to redefine the field and take it to new heights. Service computing has so far been confined to the realm of wired computing. Its real potential and great impact will be felt when it starts reflecting on the exponential use of mobile devices for entertainment and business purposes. High-speed wireless networks have now made it possible to broadcast applications that would have been unfeasible a few years ago. The content of this book is a building block in the realization of the dream of network blind (wired/wireless) service broadcast. Since access to broadcast channels is by far the costliest part of accessing services, the book emphasizes the coverage of access methods in various service broadcast channel organizations.

We wish to thank our respective families for their support during the write-up of this monograph. In particular, Athman wants to thank his wife, Malika, and his three sons, Zakaria, Ayoub, and Mohamed-Islam, for their support and patience during the preparation of this monograph. Xu wants to thank his wife, Amy, for her kind and wholehearted support.

Falls Church, Virginia
Canberra, Australia

Xu Yang
Athman Bouguettaya
December, 2008

Contents

Acronyms

3G/4G	Third/Fourth Generation of Mobile Standards and Technology
BPEL	Business Process Execution Language
CORBA	Common Object Request Broker Architecture
EDI	Electronic Data Interchange
EJB	Enterprise Java Beans
GIS	Geographical Information System
HTTP	HyperText Transfer Protocol
FTP	File Transfer Protocol
LAN	Local Area Network
MSS	Mobile Service Station
RMI	Remote Method Invocation
SMTP	Simple Mail Transfer Protocol
SOAP	Simple Object Access Protocol
WSDL	Web Service Definition Language
UDDI	Universal Description Discovery and Integration
UMTS	Universal Mobile Telecommunications System
URI	Uniform Resource Identifier
UUID	Universally Unique Identifier
Wi-Fi	Wireless Fidelity
WiMAX	Worldwide Interoperability for Microwave Access
XML	Extensible Markup Language

Chapter 1
Introduction

The Internet was originally invented as a technology for sharing information among computers for scientific research purposes. Early standards, such as telnet protocol [45], Simple Mail Transfer Protocol (SMTP) [67], and File Transfer Protocol (FTP) [46], further improved the ability of exchanging information through the Internet. By the end of 1990s, with the development of HyperText Transfer Protocol (HTTP) [13] and other core Web technologies, the Internet has evolved to become the medium for connecting hundreds of millions of computers and exchanging massive amount of information. Given its great capacity and huge customer base, the Internet has provided companies with new business opportunities. As a result, more and more companies have changed their business models and started using the Internet as an important means of conducting their daily businesses. The Internet has now become a commonly used medium for business activities, which are often referred to as *Electronic Commerce* or *E-commerce* [38, 27].

Enabling technologies for E-commerce have been around for almost three decades. They provide businesses with means for interacting with their peers (B2B E-commerce) and customers (B2C E-commerce). One of the early standards for fulfilling requirements of E-commerce is Electronic Data Interchange (EDI). EDI is a standard for the electronic exchange of information between entities using standard, machine-processable, structured data formats [6]. EDI has provided a means for different businesses to interact with each other. However, EDI requires applications to use standardized information. With great heterogeneity of the information on the Internet today, this becomes a severe limitation. A number of middleware technologies, such as CORBA, RMI, and EJB, have emerged since 1980s [42, 52, 51]. These technologies have provided new means for applications from different companies to communicate with each other over the networks. The great advantage of these technologies is that the communication details are transparent to users. Different business applications can interact with each other without having to worry about how the actual interaction takes place. However, all these middleware technologies require the interface between any two applications to be pre-defined. New interfaces would have to be built if a business entity wants its applications to interact with applications from different business partners. This greatly limits the usage of middleware

X. Yang and A. Bouguettaya, *Access to Mobile Services*, Advances in Database Systems 38, DOI: 10.1007/978-0-387-88755-5_1, © Springer Science + Business Media, LLC 2009

technologies on the Internet, especially for businesses that require the interaction to be dynamic. By *dynamic interaction*, we mean any application can dynamically choose whom to interact with based on its best business interests. For example, assume a company that assembles and sells desktop computers has an automatic ordering system to order computer parts from different vendors and each vendor also has a automated system to process orders. To find out the lowest price of a computer part, the automatic ordering system would need to query each vendor to obtain the price. With the increasing number of vendors, the system would also need to automatically discover new vendors. Since the interfaces to different vendors' systems could be different, the automatic ordering system would need to know how to interact with each vendor's system. Apparently, traditional E-commerce technologies do not work efficiently with such business activities.

More recently, a number of new technologies such as XML, WSDL (*Web Services Description Language*) [56, 59], UDDI (*Universal Description, Discovery and Integration*) [61], and SOAP(*Simple Object Access Protocol*) [57] have been developed to enable the description, discovery, and communication of Web applications in a more flexible way. This new type of applications, called *Web services*, provides a loosely coupled method for businesses to interact with each other [12]. This method makes it possible for business applications to be discovered and interact with each other dynamically [44]. A Web service is a software application identified by a URI (*Uniform Resource Identifier*) [5], whose interfaces and binding are defined, described and discovered by XML artifacts, and supports direct interactions with other software applications using XML based messages via Internet-based protocols [60]. Basic Web services infrastructure consists of *service requesters* (e.g. mobile users), *service providers* (e.g. companies providing services), and *discovery agency* (e.g. UDDI registry) [58]. Service providers register their Web services with the discovery agency. Service requesters access the discovery agency first to discover the Web services of their interest. They then obtain the descriptions (in WSDL) of the Web services to find out where they are located and how to invoke them. At last they invoke the actual Web services by using the information provided in the service descriptions. Figure 1.1 illustrates a basic Web services architecture.

Considering the automatic ordering system example again, assume the automatic ordering system and the vendors' order processing systems are all Web services. Before ordering certain computer parts, the automatic ordering system could search the discovery agency for services that provide these parts first. After obtaining the invocation details of these services, the system could then query them to find out which service provides the lowest prices for the required parts and then place an order with this service. Obviously, Web services have made dynamic interaction between business applications much easier. At present, examples of Web services span various application domains including stock trading, credit checking, real-time traffic reports, and language translation. It is expected that the number and types of Web services will increase at a fast pace in the near future [7]. Almost every asset on the Web is expected to be turned into a service that would drive new revenue streams and create new efficiencies.

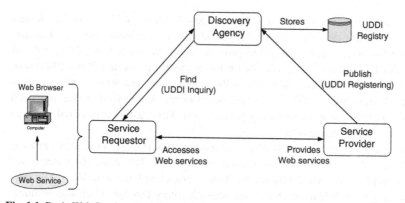

Fig. 1.1 Basic Web Services Architecture

The past years have witnessed a boom in wireless technologies [49]. Sophisticated wireless devices such as cellular phones and PDAs (*Personal Digital Assistants*) are now available at affordable prices. Emerging technologies including 3G and 4G (third and fourth generation) are under development to increase the bandwidth of wireless channels [15]. With these wireless technologies, mobile users can acquire information while on the move. This gives users great flexibility and convenience. There are increasing number of wireless applications. For example, in *Geographical Information Systems (GIS)*, mobile users could ask for geographical information to find a restaurant of their choice in the vicinity or the highest peak in the area of interest. Another example is *wireless stock market data delivery*. Stock information from any stock exchange in the world could be broadcast on wireless channels or sent to mobile users upon requests. A typical wireless system consists of wireless servers called *Mobile Service Stations (MSS)* and mobile clients, which are also referred to as *Mobile Units (MU)*. Each wireless server is equipped with wireless transmitters capable of reaching thousands of moC bile clients residing in *cells*. A cell is the geographic area covered by a wireless server. Mobile clients can move between cells and query wireless servers for information. Wireless servers are often connected to the Internet to provide mobile users with the information they need. With wireless technologies, companies can now better serve users who are frequently on the move and do not have physical access to the Internet.

Driven by the success of E-commerce and impressive progress in wireless technologies, *Mobile commerce (M-commerce)* is rapidly taking shape. M-commerce refers to the conduct of business over wireless communications and devices [64]. Examples of M-commerce applications include mobile office (e.g., working while on the move), mobile advertising (e.g., location sensitive advertisements), and mobile financial applications (e.g., banking and payment for mobile users) [63, 36]. To support wireless-oriented services in M-commerce, a new generation of Web services called *Mobile services (M-services)* has emerged. An M-service is a Web service that is accessible by mobile clients through wireless networks. M-services promise several benefits compared with their wired counterparts. First of all, M-services cater for "*anytime and anywhere*" access to services. Users need no longer sit in front of their desktop computers to conduct their business activities. Furthermore, the M-services are expected to provide larger customer base. As of the end of year 2001, there are more than 440 million mobile users and 280 million mobile consumers worldwide. More than 130 million users and 80 million customers are in the United States [47]. Analysts expect this number to further rise to exceed the total number of wired computing devices [65]. This customer base will provide M-services a huge business market.

There are two basic methods of implementing M-services, (1) *M-services over the Web* and *M-services over wireless channels* [33, 34, 35]. With M-services over the Web, mobile users wirelessly access Web services residing on the Web. All interactions between mobile users and M-services take place through wireless communications and all computations are conducted on the server side (i.e. service providers). With M-services over wireless channels, mobile users retrieve M-services from wireless channels to their terminals and execute the applications locally. M-services over the Web usually works better for applications that do not require extensive user interactions because of the often limited wireless bandwidth in practice. Furthermore, disconnect is a common problem in wireless world, which may cause a transaction to be interrupted and all its intermediate information to be lost. While M-services over wireless channels does not have this problem because all interactions and computations are performed on users' local terminals. However, sending M-services over wireless channels could be expensive. Therefore, M-services over wireless channels normally requires the applications to be lightweight. In this book, we focus on the method of M-services over wireless channels. We assume M-services are executed on mobile users's local terminals. But how do M-services get delivered to mobile users in the first place? Similar to delivering wireless data, two basic modes could be used, *broadcast/push* and *on-demand/pull*. With broadcast mode, M-services are broadcast to users and with on-demand mode, they are delivered to users upon requests. The advantages of the broadcast mode include its simplicity in implementation (e.g., requires no interaction), better scalability (e.g. supports any number of users) and little limitation on mobile clients (e.g. only requires receiving capability). Given its great advantages, the broadcast mode is often considered to be the most suitable wireless data delivery method in many scenarios, including weather forecast, traffic report, stock market report, etc. The broadcast mode works particularly well for applications that have a large user base and comparatively small set of data to be delivered. For example, it would be extremely inefficient for the wireless base stations to process a large number of requests that ask for the same service on an on-demand basis. A more suitable approach for this scenario is broadcasting. Frequently accessed services can be made available on broadcast channels so that mobile users can retrieve them directly. We refer to such services as broadcast based M-services, which normally have the following characteristics:

- The provided services are specific to a geographic region, such as a city of a metropolitan area.

- The services are usually used to retrieve real-time information, such as traffic reports, weather forecasts, event schedules, availability for movie tickets, etc.
- The services are information oriented, which means they do not require much user interaction. Users may retrieve the information based on certain filtering criteria. For example, a user could use a service to look for a French restaurant nearby that still has seats available.
- The services and wireless data are frequently accessed.
- The services and wireless data are accessed by a large number of users.

In this book, we assume M-services are lightweight and consider broadcast to be the main method of delivering M-services to mobile users. A few examples for broadcast based M-service systems are as follows:

- *Regional M-service system* which normally covers the range of a town or city. The provided services may include live traffic reports, stock market updates, weather forecasts, and etc.
- *Campus M-service system* which covers the range of a university campus. The provided services may include student activities, class schedules, parking, shuttle schedules, and etc.
- *Auction M-service system* which covers the range of one or more buildings that hold an auction. The provided services may include action items catalog, auction status report, bidding service, and etc.

In this book, we propose an M-services infrastructure which provides a generic framework for mobile users to lookup, access and execute Web services over wireless broadcast networks. One of the most important issues in wireless broadcast networks is efficient access to broadcast data [23]. The access efficiency is usually measured by two factors: *client waiting time* and *power consumption by mobile devices*. The second factor is of particular interest due to the fact that most mobile devices are often used away from fixed power source and equipped with limited power supply. These factors are normally measured by the following two parameters:

- *Access Time*: The average time that has elapsed from the moment a client requests information up to the point when the required information is downloaded by the client. This factor corresponds to client waiting time.
- *Tuning Time*: The amount of time that has been spent by a client listening to broadcast channels. This is used to determine the power consumed by the client to retrieve the required information.

In recent years, a lot of research work has been conducted in designing efficient data access methods for push based wireless networks [26, 25, 29, 54, 32, 53, 17, 28]. These data access methods are used to help mobile users locate requested data in broadcast channels more efficiently. However, these methods cannot be directly applied to the M-service environment because accessing an M-service is inherently different from accessing wireless data. First of all, accessing services is a different process from accessing wireless data. Accessing data is usually considered as a simple "search and match" action. Accessing services, on the other hand, requires multiple steps including service discovery, service downloading, service execution, and data retrieval. Furthermore, a service itself could be dependent on other services. This may requires access methods to download and execute all depended services first. In this book, we propose novel access methods and broadcast channel organizations for mobile users to efficiently access M-services and wireless data in wireless broadcast networks. In particular, we focus on discussing efficient access to composite M-services, which can be depicted by *Business Process Execution Language (BPEL)* [30]. We only consider *pre-defined* composite M-services in this book. By *pre-defined*, we mean the BPEL definitions of all supported composite M-services are already generated and available to mobile users.

The rest of the book is organized as follows:

- Chapter 2 provides an overview of broadcast based M-services and discusses issues and challenges of efficient access to M-services.
- Chapter 3 conducts analytical and practical study on different access methods in traditional wireless data networks.
- Chapter 4 proposes new methods for further improving data access efficiency in wireless data networks.
- Chapter 5 discusses accessing simple M-services using existing access methods.
- Chapter 6 defines semantics for accessing composite M-services and studies how to leverage these semantics to achieve best access efficiency.
- Chapter 7 presents a few effective channel organizations for delivering composite M-services and wireless data to mobile users.
- Chapter 8 presents practical study on the proposed access methods and channel organizations.
- Chapter 9 discusses open problems for accessing M-services.

Chapter 2
Access to Broadcast M-services: Issues and Challenges

In this chapter, we discuss the need for new access methods and an accompanying infrastructure for broadcast based M-services. The infrastructure of traditional wireless broadcast systems cannot keep up with the rapid development of wireless technologies and the fast increasing wireless services. It is getting more difficult for mobile users to find suitable wireless services. Existing infrastructure also does not facilitate collaboration of multiple services. In this book, we discuss applying Web services technologies into wireless broadcast environments and propose a new M-services infrastructure. One of the most important issues in broadcast environments is efficient access to broadcast information. A wide range of work has been done in the past two decades in studying efficient access in traditional data wireless broadcast systems. However, new access methods are needed due to the following new challenges for broadcast M-services systems:

- *Access pattern*: Accessing M-services is a different process from accessing data and has a different access pattern.
- *Service dependencies*: Services could have various dependencies between each other that could affect access efficiency.
- *Service-data dependencies*: Each service could request for one or more data items. Access methods should consider the dependencies between services and their required data.
- *Access semantics*: There are a few semantics that have impact on access efficiency and should be considered by access methods.

The focus of this book is to investigate efficient access methods for broadcast based M-services. The new access methods should address the challenges mentioned above.

2.1 Broadcast based M-services

Recent wireless broadband technologies, such as Wi-Fi, WiMaX, UWB and 3G/UMTS are bringing the promise of large bandwidth everywhere. Wireless applications today enjoy much larger bandwidth than in the past. Wireless bandwidth is expected to be further increased with emerging technologies in the near future. With the extra bandwidth at hand, we now can deploy more wireless applications and make more information available on the air. However, with the increasing number of wireless applications and larger amount of wireless data, it would be difficult for mobile users to discover new wireless applications using the subscribing approach used in the existing wireless broadcast networks. Furthermore, In a wireless network, wireless applications are normally executed independently. With the increasing number and variety of wireless services, it is possible that a complex user request may require multiple applications. It is also possible that a wireless application may need to invoke other applications. In these cases, we require multiple applications work together to fulfill user requests. The current wireless broadcast infrastructure provides no capability for such operations.

Let us now illustrate the problem using an example. Assume professor Edward comes to New York city to attend a conference and he decides to stay for three more days after the conference just to enjoy the city a little more. Professor Edward has a to-do list, which contains the things he would like to do during his stay. Given this list and his personal preferences (such as favorite food), professor Edward would like to know the best schedule to enjoy his three days in New York city. Assume there is a broadcast based wireless information system that covers the city of New York. Professor Edward has a wireless device that could be used to access the system. He hopes to use the information system to help schedule his stay. However, with the traditional wireless broadcast infrastructure, it is difficult for the professor to take full advantage of the system. First of all, since the system has no support for service discovery, the mobile device does not know what services to use to obtain required information. Also, the system does not facilitate collaboration of multiple services. Therefore, the professor would not be able to use the system to achieve complex requests such as scheduling tasks.

In this book, we propose a new infrastructure to address these issues for traditional wireless broadcast systems. The proposed infrastructure applies the Web services technologies to wireless world. The registry and discovery technologies help mobile users find services of their interest. The service composition technologies can be used to define services that are dependent on other services. Similar to "wired" Web services, a basic wireless Web service (i.e. M-service) architecture would also consist of a service requester, a service provider and a discovery agency. M-services are made available to mobile users through wireless channels. Mobile users first look up specific M-services in a wireless service registry (discovery agency). The users then access the selected M-services through wireless channels by using the information obtained from the registry.

Figure 2.1 shows a typical example of M-services systems. A service provider advertises its services by registering the description of these services with a wireless

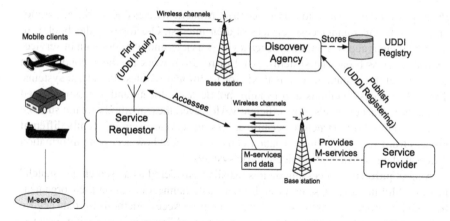

Fig. 2.1 A Typical M-service System

discovery agency. Similar to WSDL in Web services, the description contains information on how to access and invoke the services. The wireless discovery agency would make the service description available to mobile users on the wireless channels. These wireless channels are maintained by base transceiver stations. By accessing discovery agency's wireless channels, mobile clients would find services that best meet users' needs, and get information on how to access these services. The actual services are also made available through wireless channels. Mobile clients could then retrieve the services and invoke their operations. Although Web services technologies are widely used, the challenge is how to support them in broadcast-based wireless environment. The proposed infrastructure covers the following aspects:

- **Broadcast content**: The infrastructure defines what information needs to be available on broadcast channels to support broadcast-based M-services.
- **Channel layout**: The infrastructure defines different channel types and their organizations for delivering the defined broadcast content.
- **Service/data retrieval**: The infrastructure also defines how to retrieve services and wireless from broadcast channels.

2.2 Efficient Access to M-services and Wireless Data

Traditional wireless broadcast systems are normally data-centric. Mobile devices use pre-installed applications to access wireless data from broadcast channels. Stimulated by the recent success in wireless technologies, the number of mobile users have been rapidly increasing over the past years. Wireless networks are capable of delivering much larger amount of data to mobile users. It is getting more and more difficult to support fast increasing wireless data and applications with the traditional data-centric infrastructure. As already discussed, no effective way is available for mobile users to discover or combine services. As a result, wireless paradigm is shifting from data-centric to service-centric. Modern wireless networks allow mobile users discover and access services of their interest. Our focus in this book is to design efficient access methods for retrieving services and wireless data in service-centric wireless broadcast systems. A lot of research work has been conducted in designing efficient data access methods for traditional wireless broadcast systems. These data access methods aim to help mobile users locate and access requested data in broadcast channels more efficiently. However, these methods are not suitable for the proposed infrastructure because accessing services is an inherently different process from accessing wireless data. Let us now look at how accessing information differs in data-centric and service-centric systems.

Access pattern – Accessing data is usually considered as a "search and match" process. Mobile devices scan through broadcast channels based on their organizations and filtering criteria to locate requested items. Access methods for data centric systems only need to consider this "search and match" process. Accessing services, on the other hand, is a more complex process. It requires multiple steps. Users first need to discover suitable services. Then the selected services needs to be downloaded and executed. At last wireless data required by these services is retrieved. Access methods for service-centric systems need to take all these steps into consideration and try to make the whole process most efficient instead of just one single step.

Service dependencies – There are two types of services, *simple* and *composite*. Composite services may need to invoke other services. Based on their BPEL definitions, these services could have certain dependencies between each other. This means services can only be executed conforming to the defined dependencies. Therefore, it is important to know the type of a service and its defined dependencies to access and execute the service. Access methods for service-centric systems need to leverage the knowledge of these dependencies to achieve the best possible overall access efficiency.

Service-data dependencies – When a service is executed, it might need to retrieve wireless data from broadcast channels. The retrieved data could be presented to users as part of final results or used by other services as intermediate results. The required wireless data will not be known until the service is executed. For composite services, access methods may need to dynamically adjust the access sequence based on the locations of the required wireless data to achieve better performance. For example, assume composite service S defines two parallel simple services $S1$ and $S2$. Based on the locations of these services, a mobile device is to access $S2$ right after $S1$. When $S1$ is executed, it asks for data $D1$. Based on the mobile device realizes that $D1$ will arrive before $S2$. The mobile device then adjusts the pre-defined access sequence and retrieves $D1$ before $S2$.

Access semantics – Services and wireless data are delivered on a "push" instead of "pull" basis. This means some services might arrive when their dependencies are not fulfilled yet. On some mobile devices, based on available resources, they could be cached to be executed later. Alternatively, they would have to be downloaded on their next arrivals. It is important to understand these semantics and leverage such knowledge to achieve the best possible access efficiency. In this book, we define the semantics that have impact on access efficiency. Then we investigate efficient methods for accessing services taking these semantics into account.

2.3 Example Scenario

Let us use the same example presented in Section 2.1 to illustrate the use of broadcast based M-services. Assume the information system supports broadcast based M-services. and there is a service called `trip_planner`. The BPEL design of the service is shown in Figure 2.2, which is generated using Oracle JDeveloper BPEL Designer 10g. This service is a composite service that uses a few other services to schedule trips for tourists based on their input. This service first invokes other required services to gather information and then calculates the best schedule.

As shown in the BPEL graph, the `trip_planner` service invokes six other services. First, it invokes `attractions`, `entertainment`, `restaurants`, and `shopping` services to let a user select activities for this trip. These services could be accessed and executed in parallel. Once the user makes the selections, the service would invoke `transportation`, `traffic`, and `weather` services to generate the best possible schedule. The schedule is generated based on the combination of several factors, such as the time and locations of activities, and weather conditions (e.g. outdoor activities are preferred on sunny days). The best transportation between any two consecutive activities would also take traffic conditions into consideration. This makes the `traffic` service to be dependent on the `transportation` service. The `trip_planner` service could be re-executed anytime during the trip to update the existing schedule in the case of any unexpected events or change of the activities.

This example scenario shows that one user request may result in retrieving multiple services. These services could be accessed and executed in different fashions (e.g. in sequence or parallel). Similarly, wireless data could also be retrieved in various ways by different services. Access methods for such an M-services system should support the variety of ways of efficiently accessing services and data based on different user requirements.

2.4 Main Contributions of This Book

In this section, we present the overview of our main contributions in this book. We first conduct an extensive study on a few existing access techniques for data-oriented wireless networks. Then we propose a few new data access methods that outperform the existing methods. We also develop a testbed for evaluating these data access methods. We conduct extensive experiments for comparing the existing and new access methods. We also study the behavior of accessing simple M-services using traditional access methods. Then we propose an infrastructure that supports

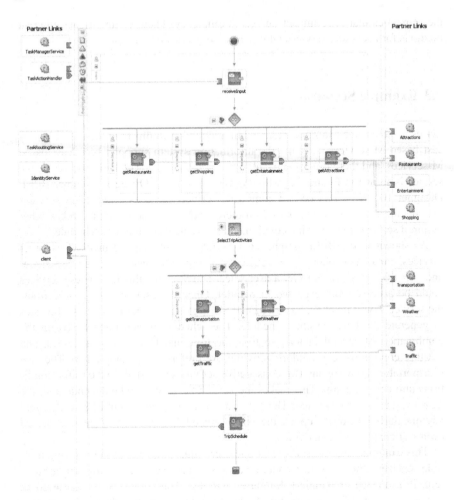

Fig. 2.2 Trip Planning Composite Service

both simple and composite services. We propose efficient access methods for the infrastructure and conduct extensive practical study on the proposed methods.

Analytical and practical study on traditional data access methods – In recent years, several wireless data access methods have been proposed to improve data access efficiency in wireless environment. These methods are evaluated in different environment contexts defined by the authors. Hence, it is difficult for readers to compare these methods because different environmental settings are used. To better analyze different access methods, we present a common analytical model for wireless environment. We define a set of commonly used environmental parameters that have impact on access efficiency. We also derive analytical cost models for a few most recognized access methods. Furthermore, a testbed has been developed to implement data access in wireless environment and simulate a wireless broadcast environment. The purpose of this testbed is to provide a platform to compare, evaluate, and help develop new access methods. The testbed supports several adjustable environmental settings for studying access methods under different settings. Extensive experiments have been conducted to compare the selected data access methods.

New data access methods – An adaptive wireless data access method and some variations are proposed for wireless environment. The proposed method is based on the observation that the performances of existing methods are severely affected by environmental settings. This proposed method exhibits good overall performance and stabler performance when some important environmental settings change. By varying a construction factor, the proposed methods would yield different performance patterns. The testbed has been enhanced to support these new access methods. Simulation experiments have been conducted to compare the proposed methods to the existing ones.

Efficient access to simple M-services using traditional techniques – We then extend our study on traditional access techniques to the M-services world. We study the behavior of accessing simple M-services using a few different traditional techniques. We assume mobile users are able to discover services of their interest using broadcast registry information. We derive analytical models for accessing simple M-services with traditional access techniques. The testbed is further enhanced to support accessing simple M-services. By using the testbed, we conduct practical study on accessing simple M-services using traditional access techniques.

Broadcast-based M-services infrastructure – We propose a novel wireless broadcast infrastructure that supports discovery and composition of M-services. We define system roles for the infrastructure and present an interaction model which defines how these system roles interact with each other. We also define the broadcast content for this infrastructure. The broadcast content contains all information mobile users need to discover, download, and execute simple and composite services.

Semantic access to broadcast-based M-services – As already discussed, accessing M-services especially composite M-services is different from accessing static wireless data. In this book, we define semantics of accessing composite M-services and study how to leverage these semantics to achieve best possible access performance. We also propose a few channel organizations for efficiently accessing services and wireless data in the proposed infrastructure. We derive analytical models for the proposed channel organizations.

Practical study on proposed access methods – A new testbed is developed to simulate the proposed M-services infrastructure. The testbed supports the defined access semantics for composite M-services. All proposed access methods and channel organizations are implemented in the testbed. We conduct extensive experiments to study the access efficiency of the proposed access methods and channel organizations. We also study the impact of different semantics on these methods.

Chapter 3
Traditional Data Access Methods

Access efficiency has drawn a lot of research attention in the past few decades. Several wireless data access methods have been proposed to improve data access efficiency in wireless environment. These methods are evaluated in different environment contexts defined by the authors. Hence, it is difficult for readers to compare these methods because different environmental settings are used. To better analyze different access methods, we present a common analytical model for wireless environment. We define a set of commonly used environmental parameters that have impact on access efficiency. We selected a few most recognized access methods and derived analytical cost models for them. A testbed is developed to implement data access in wireless environment and simulate a wireless broadcast environment. The purpose of this testbed is to provide a platform to compare, evaluate, and help develop new access methods. The testbed supports several adjustable environmental settings for studying access methods under different settings. Extensive experiments have been conducted to compare the selected data access methods.

3.1 Basic Data Access Techniques

In recent years, several data access methods have been proposed to improve performance of data access by introducing indexing techniques to broadcast based wireless environments. Most of these methods are based on three basic techniques: *index tree* [24, 8, 50, 22, 48, 21, 55, 16], *signature indexing* [29, 9], and *hashing* [25]. Some methods are based on the combination of more than one of these techniques. For example, indexing methods taking advantages of both index tree and signature indexing techniques have been proposed in the past [54, 18, 19, 20]. In the rest of this section, we discuss these three basic techniques in details.

X. Yang and A. Bouguettaya, *Access to Mobile Services*, Advances in Database Systems 38, DOI: 10.1007/978-0-387-88755-5_3, © Springer Science + Business Media, LLC 2009

3.1.1 Index tree based access methods

B+ tree indexing is a widely used indexing technique in traditional disk-based environments. It is also one of the first indexing techniques applied to wireless environments. The use of B+ tree indexing in wireless environments is very similar to that of traditional disk based environments. Indices are organized in B+ tree structure to accelerate the search processes. An offset value is stored in each index node pointing at corresponding data item or lower level index node. However, there are some differences that introduce new challenges to wireless environments. For example, in disk based environments, offset value is the location of the data item on disk, whereas in wireless environments, offset value is the arrival time of the bucket containing the data item. A *bucket* is the basic logical unit of a broadcast channel, which usually contains a data item or several index entries with other access method specific information. Moreover, indices and data in wireless environments are organized in one-dimensional mode in broadcast channel. Missing the bucket containing index of the requested data item may cause the client to wait until the next broadcast cycle to find it again. The most representative B+ tree based methods are *(1,m) indexing* and *distributed indexing* [24]. We only discuss distributed indexing in this section, because it is derived from (1,m) indexing and they have very similar data structure.

3.1.1.1 Data Organization

In distributed indexing, every broadcast data item is indexed on its primary key attribute. Indices are organized in B+ tree structure. Figure 3.1 shows a typical full index tree consisting of 81 data items [24]. Each index node has a number of pointers (in Figure 3.1, each node has three pointers) pointing at its child nodes. The pointers of the bottom level indices point at the actual data nodes. To find a specific data item, the search follows a top-down manner. The top level index node is searched first to determine which child node contains the data item. Then the same process will be performed on that node. This procedure continues till it finally reaches the data item at the bottom. The sequence of the index nodes traversed is called the *index path* of the data item. For example, the index path of data item 34 in Figure 3.1 is I, a2, b4, c12.

What was discussed so far is similar to the traditional disk-based B+ tree indexing technique. The difference arises when the index and data are put in the broadcast channel. A node in the index tree is represented by an index bucket in the broadcast channel. Similarly, a broadcast data item is represented by a data bucket. In the traditional disk-based systems, index and data are usually stored in different locations. The index tree is searched first to obtain the exact location of the requested data item. This process often requires frequent shifts between index nodes or between index and data nodes. As data in a wireless channel is one-dimensional, this kind of shift is difficult to achieve. Therefore, in distributed indexing, data and index are interleaved in the broadcast channel. The broadcast data is partitioned into several

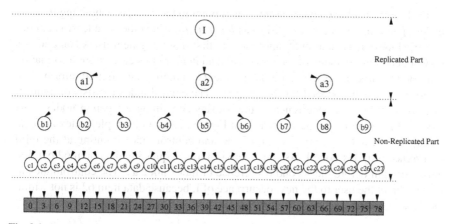

Fig. 3.1 A sample index tree

data segments. The index tree precedes each data segment in the broadcast. Users traverse the index tree first to obtain the time offset of the requested data item. They then switch to doze mode until the data item arrives. Figure 3.2 illustrates how index and data are organized in the broadcast channel.

In (1,m) indexing [24], the whole index tree precedes each data segment in the broadcast. Each index bucket is broadcast a number of times equal to the number of data segments. This increases the broadcast cycle and thus access time. Distributed indexing achieves better access time by broadcasting only part of the index tree preceding each data segment. The whole index tree is partitioned into two part: replicated part and non-replicated part. Every replicated index bucket is broadcast before the first occurrence of each of its child. Thus the number of times it is broadcast is equal to the number of children it has. Every non-replicated index node is broadcast exactly once, preceding the data segment containing the corresponding data records. Using the index tree in Figure 3.1 as an example, the first and second index segments will consist of index buckets containing nodes I, a1, b1, c1, c2, c3 and a1, b2, c4, c5, c6 respectively.

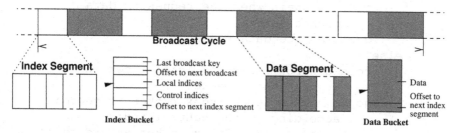

Fig. 3.2 Index and data organization of distributed indexing

Each index bucket contains pointers that point to the buckets containing its child nodes. These pointers are referred to as *local index*. Since the broadcast is continuous and users may tune in at any time, the first index segment users come across may not contain the index of the requested data item. In this case, more information is needed to direct users to other index segment containing the required information. *Control index* is introduced for this purpose. The control index consists of pointers that point at the next occurrence of the buckets containing the parent nodes in its index path. Again using the index tree in Figure 3.1 as an example, index node a2 contains local index pointing at b4, b5, b6 and control index pointing at the third occurrence of index node I. Assume there is a user requesting data item 62 but first tuning in right before the second occurrence of index node a2. The control index in a2 will direct the user to the next occurrence of I, because data item 62 is not within the subtree rooted at a2.

3.1.1.2 Access Protocol

The following is the access protocol of distributed indexing for a data item with key K:

```
        mobile client requires data item with key K
        tune into the broadcast channel
        keep listening until the first complete bucket
            arrives
        read the first complete bucket
        go to the next index segment according to the
            offset value in the first bucket
   (1)  read the index bucket
        if K < Key most recently broadcast
            go to next broadcast
        else if K = Key being broadcast
            read the time offset to the actual data
                records
            go into doze mode
            tune in again when the requested data
                bucket comes
            download the data bucket
        else
            read control index and local index in
                current index bucket
            go to higher level index bucket if needed
                according to the control index
            go to lower level index bucket according to
                the local index
            go into doze mode between any two successive
                index probes
```

```
repeat from (1)
```

3.1.2 Signature indexing

A signature is essentially an abstraction of the information stored in a record. It is generated by a specific signature function. By examining a record's signature, one can tell if the record possibly has the matching information. Since the size of a signature is much smaller than that of the data record itself, it is considerably more power efficient to examine signatures first instead of simply searching through all data records. Access methods making use of signatures of data records are called *signature indexing*. Three signature indexing based access methods, *simple signature, integrated signature,* and *multi-level signature*, are proposed in [29]. The integrated and multi-level signature indexing methods are based on the simple signature indexing method and designed to handle more complex data structures. We only discuss the simple signature method in this section.

The signatures are generated based on all attributes of data records. A signature is formed by hashing each field of a record into a random bit string and then superimposing together all the bit strings into a record signature. The number of collisions depends on how perfect the hashing function is and how many attributes a record has. Collisions in signature indexing occur when two or more data records have the same signature. Usually the more attributes each record has, the more likely collisions will occur. Such collisions would translate into false drops. False drops are situations where clients download the wrong data records which happen to have matching signatures.

3.1.2.1 Data Organization

In signature based access methods, signatures are broadcast together with data records. The broadcast channel consists of index (signature) buckets and data buckets. Each broadcast of a data bucket is preceded by a broadcast of the signature bucket, which contains the signature of the data record. For consistency, signature buckets have equal length. Mobile clients must sift through each broadcast bucket until the required information is found. The data organization of simple signature indexing is illustrated in Figure 3.3.

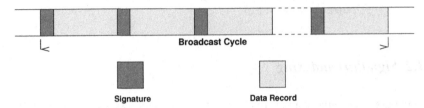

Fig. 3.3 Data organization of signature indexing

3.1.2.2 Access Protocol

The access protocol for simple signature indexing is as follows (assume K and S are the key and signature of the required record respectively, and $K(i)$ and $S(i)$ are the key and signature of the i-th record):

```
        mobile client requires data item with key K
        tune in to broadcast channel
        keep listening until the first complete signature
            bucket arrives
   (1)  read the current signature bucket
        if S(i) = S(k)
            download the data bucket that follows it
            if K(i) = K
                search terminated successfully
            else
                false drop occurs
                continue to read the next signature bucket
                repeat from (1)
        else
            go to doze mode
            tune in again when the next signature bucket
                comes
            repeat from (1)
```

3.1.3 Hashing

Hashing is another well-known data access technique for traditional database systems. In this subsection, we introduce a *simple hashing* method, which was proposed for broadcast based wireless environments [25].

3.1.3.1 Data Organization

Simple hashing method stores hashing parameters in data buckets without requiring separate index buckets or segments. Each data bucket consists of two parts: *Control part* and *Data part*. The *Data part* contains actual data record and the *Control part* is used to guide clients to the right data bucket. The data organization of the broadcast channel using simple hashing is illustrated in Figure 3.4.

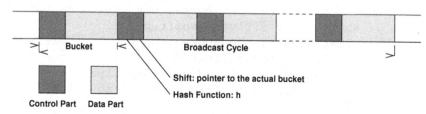

Fig. 3.4 Index and data organization of simple hashing

The control part of each data bucket consists of a *hashing function* and a *shift value*. The hashing function maps the key value of the data in the broadcast data record into a hashing value. Each bucket has a hashing value *H* assigned to it. In the event of a collision, the colliding record is inserted right after the bucket which has the same hashing value. This will cause the rest of records to shift, resulting in data records being "out-of-place". The shift value in each bucket is used to find the right position of the corresponding data record. It points to the first bucket containing the data record with the right hashing value. Assume the initial allocated number of buckets is N_a. Because of collisions, the resulting length of the broadcast cycle after inserting all the data records will be greater than N_a. The control part of each of the first N_a buckets contains a shift value (offset to the bucket containing the actual data record with the right hashing value), and the control part of each of the remaining data buckets contains an offset to the beginning of the next broadcast.

3.1.3.2 Access Protocol

Assume the hashing function is *H*, thus the hashing value of key *K* will be *H(K)*. The access protocol of hashing for a data item with key *K* is:

```
      mobile client requires data item with key K
      tune into the broadcast channel
      keep listening until the first complete bucket
          arrives
(1)   read the bucket and get the hashing value h
      if h < H(K)
          go to doze mode
          tune in again when h = H(K)  (hashing
```

```
                      position)
         else
             go to doze mode
             tune in again at the beginning of the next
                 broadcast
             repeat from (1)
         read shift value at the bucket where h = H(K)
         go to doze mode
         tune in again when the bucket designated by the
             shift value arrives (shift position)
         keep listening to the subsequent buckets, till
             the wanted record is found
                 search terminated successfully
             or a bucket with different hashing value
             arrives
                 search failed
```

3.2 Analytical Study

Access Time and *Tuning Time* are the two factors that are commonly used to measure the efficiency of wireless data access methods. The access time refers to the total time mobile clients need to wait for the request to complete. The tuning time is the actual time spent by mobile clients to actively listen to wireless channels and process requests. Request processing requires CPUs to stay busy. Listening to wireless channels means that receiving devices are actively retrieving data from wireless channels. Since most power consuming parts of a mobile unit are CPU and receiving devices [66], the tuning time of a request is usually proportional to the power consumed by a mobile unit on the request.

Data Access Methods are used to improve the efficiency (usually the power consumption) of wireless data access. Several data access methods have been proposed in recent years to conserve power consumption in wireless environments. Each of them has its own advantages and drawbacks. However, since these methods are presented with different environmental settings, it is difficult to compare them in quantitized manner. In this section, we first define a basic wireless environment that provides mobile users with information through wireless broadcast channels. The environment will be served as unified context to evaluate various access methods. Then we present the analytical evaluation models for the selected access methods under the unified environment.

3.2.1 Basic Broadcast-based Wireless Environment

In the basic broadcast-based wireless environment, we assume there is only one broadcast channel because most of the existing wireless data access methods are proposed for single channel scenario. A mobile user obtains the required information by listening to the broadcast channel till the data of interest is broadcast and downloaded to the mobile client. We define the following parameters for the environment:

System parameters	
N_r	Number of broadcast data items
N	Number of total buckets
S_{dk}	Key size of data items
S_d	Data item size
S_b	Logical broadcast unit (bucket) size
B_c	Broadcast cycle - the length of all contents in the broadcast channel
B_d	Broadcast channel bandwidth
Performance measurement parameters	
A_t	Access time
T_t	Tuning time
B_t	Broadcast cycle time - time to scan the whole broadcast channel
I_t	Time to browse an index bucket
D_t	Time to browse a data bucket
F_t	Time to reach the first complete bucket (initial wait)

Table 3.1 Symbols and parameters for data access methods

Since the time that a user start listening is totally random, the mobile client may hit in the middle of a broadcast bucket after tuning into the broadcast channel. The initial wait time is the time spent to reach the first complete bucket after tuning into the broadcast channel. The *number of broadcast data items* (N_r), represents the number of data items being broadcast in a broadcast cycle. The *number of total buckets* (N) designates the total number of buckets in a broadcast cycle, including data buckets (containing data items), index buckets (containing indices in index tree based methods), and hashing buckets (containing hash values in hashing based methods).

When there is no access method (flat broadcast), users must keep listening to the broadcast channel until the required data item arrives. Therefore, the average access time and tuning time are half of the whole broadcast cycle plus the initial wait time, which can be expressed as follows:

$$A_t = T_t = F_t + B_t$$
$$= (\frac{1}{2} + N_r) \times D_t$$

3.2.2 Cost model for index tree based access methods

We now derive the access and tuning times for index tree based access methods. First, we define symbols which are specific to these methods. Let n be the number of indices contained in an index bucket, let k be the number of levels of the index tree, and let r be the number of replicated levels. It is obvious that $k = \lceil \log_n(N_r) \rceil$. The access time consists of three parts: *initial wait*, *initial index probe*, and *broadcast wait*.

initial wait (F_t): It is the time spent to reach the first complete bucket. Obviously we have:

$$F_t = \frac{D_t}{2}$$

initial index probe (P_t): This part is the time to reach the first index segment. It can be expressed as the average time to reach the next index segment, which is calculated as the sum of the average length of index segments and data segments. Given the number of replicated level is r, the number of replicated index (N_{rp}) is:

$$1 + n + \ldots + n^{r-1} = \frac{n^r - 1}{n - 1}$$

The number of non-replicated index (N_{nr}) is:

$$n^r + n^{r+1} + \ldots n^{k-1} = \frac{n^k - n^r}{n - 1}$$

As we mentioned before, each replicated index is broadcast n times and each non-replicated index is broadcast exactly once. Thus the total number of index buckets can be calculated as:

$$N_{rp} \times n + N_{nr} = n \times \frac{n^r - 1}{n - 1} + \frac{n^k - n^r}{n - 1} = \frac{n^k + n^{r+1} - n^r - n}{n - 1}$$

The number of data segments is n^r because the replicated level is r. Thus the average number of index buckets in an index segment is:

$$\frac{1}{n^r} \times (n \times \frac{n^r - 1}{n - 1} + \frac{n^k - n^r}{n - 1}) = \frac{n^{k-r} - 1}{n - 1} + \frac{n^{r+1} - n}{n^{r+1} - n^r}$$

The average number of data buckets in a data segment is $\frac{N_r}{n^r}$. Therefore, the initial index probe is calculated as:

$$P_t = \frac{1}{2} \times (\frac{n^{k-r} - 1}{n - 1} + \frac{n^{r+1} - n}{n^{r+1} - n^r} + \frac{N_r}{n^r}) \times D_t$$

broadcast wait (W_t): This is the time from reaching the first indexing segment to finding the requested data item. It is approximately half of the whole broadcast cycle, which is $\frac{N}{2} \times D_t$. Thus, the total access time is:

$$A_t = F_t + P_t + W_t$$

$$= \frac{1}{2} \times \left(\frac{n^{k-r} - 1}{n - 1} + \frac{n^{r+1} - n}{n^{r+1} - n^r} + \frac{N_r}{n^r} + N + 1 \right) \times D_t$$

The tuning time is much easier to calculate than access time, because during most of the probes clients are in doze mode. The tuning time includes the initial wait ($\frac{D_t}{2}$), reading the first bucket to find the first index segment (D_t), reading the control index to find the segment containing the index information of the requested data item (D_t), traversing the index tree ($k \times D_t$), and downloading the data item (D_t). Thus, the tuning time is:

$$T_t = \left(k + 3\frac{1}{2}\right) \times D_t$$

3.2.3 Cost model for signature indexing

For signature indexing, clients must scan buckets one by one to find the required information, the access time is determined by the broadcast cycle. A signature bucket contains only the signature of a data record. No extra offset or pointer value is inserted into the signature/index bucket as in other access methods. Since the total length of all data records is a constant, the length of signatures is the only factor that determines the broadcast cycle. Access time, therefore, is determined by the length of signature buckets. The smaller the signatures are, the better the access time is. As for tuning time, it is determined by two factors: the size of signature buckets and the number of false drops. It is obvious that smaller signature lengths reduce tuning time. However, smaller signature sizes usually implies more collisions or false drops. In cases of false drops, wrong data records are downloaded by mobile clients, resulting in longer tuning time. From this analysis, we observe two trade-offs: (1) signature length against tuning time, and (2) access time against tuning time.

Signature indexing uses two types of buckets (with varying sizes): signature bucket and data bucket. The initial wait is the time to reach the closest signature bucket:

$$F_t = \frac{1}{2} \times (D_t + I_t)$$

As discussed above, the access time is determined by the broadcast cycle. It consists of two parts: the initial wait (F_t) and the time to browse the signature and data buckets (SD_t). The average value of SD_t for retrieving a requested bucket is half of the broadcast cycle ($\frac{1}{2} \times (D_t + I_t) \times N_r$). Therefore, the access time is:

$$A_t = F_t + SD_t$$

$$= \frac{1}{2} \times (D_t + I_t) + \frac{1}{2} \times (D_t + I_t) \times N_r$$

$$= \frac{1}{2} \times (D_t + I_t) \times (N_r + 1)$$

The tuning time is determined by both the length of index buckets and the number of false drops. It consists of three parts: the initial wait (F_t), the time to browse signature buckets (SB_t), and the time to retrieve false drop data buckets (FD_t) and the requested data bucket (D_t). The average value of SB_t is half of the total length of signature buckets, which is $\frac{1}{2} \times I_t \times N_r$. Assuming F_d is the number of false drops, the value of FD_t will be $F_d \times D_t$. Hence the resulting tuning time is:

$$
\begin{aligned}
T_t &= F_t + SB_t + FD_t + D_t \\
&= \frac{1}{2} \times (D_t + I_t) + \frac{1}{2} \times I_t \times N_r + F_d \times D_t + D_t \\
&= \frac{1}{2} \times (N_r + 1) \times I_t + (F_d + 1\frac{1}{2}) \times D_t
\end{aligned}
$$

3.2.4 Cost model for hashing

The access time of the hashing method consists of an initial wait time (F_t), time to reach the *hashing position* (H_t), time to reach the *shift position* (S_t), time to retrieve colliding buckets (C_t), and time to download the required bucket (D_t). Since there is only one type of bucket used in hashing, the initial wait is $F_t = \frac{D_t}{2}$. Let N_c be the number of colliding buckets, the average number of shifts of each bucket is thus $\frac{N_c}{2}$. Therefore, we have $S_t = \frac{N_c}{2} \times D_t$. Furthermore, the average number of colliding buckets for each hashing value is $\frac{N_c}{N_r}$. Thus, we have $C_t = \frac{N_c}{N_r} \times D_t$. The calculation of H_t is more involved. Assume the number of initially allocated buckets is N_a. The resulting total number of buckets in the broadcast cycle is $N = N_a + N_c$. We have the following three possibilities that result in different values of H_t (assume the position of the first arriving bucket is n).

$$H_{t1} = \frac{N_c}{N} \times (\frac{1}{2} \times (N_c + N_a)) \qquad\qquad (n > N_a)$$

$$H_{t2} = \frac{1}{2} \times \frac{N_a}{N} \times \frac{N_a}{3} \qquad (n \leq N_a \text{ and } request\_item\_broadcast = False)$$

$$H_{t3} = \frac{1}{2} \times \frac{N_a}{N} \times (\frac{N_a}{3} + N_c + \frac{N_a}{3}) \quad (n \leq N_a \text{ and } request\_item\_broadcast = True)$$

The *request_item_broadcast* above designates if the requested information has already been broadcast in the current broadcast cycle. The first part of each formula above is the probability the scenario will happen. As a result, we have $H_t = H_{t1} + H_{t2} + H_{t3}$. Based on the above discussion, the access time is:

$$A_t = F_t + H_t + S_t + C_t + D_t$$
$$= F_t + H_{t1} + H_{t2} + H_{t3} + S_t + C_t + D_t$$
$$= (\frac{1}{2} + \frac{N}{N_a} + N - \frac{1}{2} \times N_a) \times D_t$$

The tuning time consists of an initial wait time (F_t), time to read the first bucket to obtain the hashing position (D_t), time to obtain the shift position (D_t), and time to retrieve the colliding buckets (C_t), and time to download the required bucket (D_t). The probability of collision is $\frac{N_c}{N_r}$. Thus, we have $C_t = \frac{N_c}{N_r} \times D_t$. For those requests that tune in at the time which the requested bucket has already been broadcast, one extra bucket read is needed to start from the beginning of the next broadcast cycle. The probability of this scenario occurrence is $(N_c + \frac{1}{2} \times N_r)/(N_c + N_r)$. As a result, the expected tuning time is:

$$T_t = (\frac{1}{2} + \frac{N_c + \frac{1}{2} \times N_r}{N_c + N_r} + \frac{N_c}{N_r} + 3) \times D_t$$

3.3 Testbed

This section presents the testbed we developed for the evaluation of data access methods in wireless environments. The testbed is implemented in Java language using the *JavaSim* simulation package [43] [31]. The testbed simulates data access for traditional wireless environments. The testbed is event-driven. The broadcasting of each data item, generation of each user request and processing of the request are all considered to be separate events in this testbed. They are handled independently without interference with each other. We call the testbed *adaptive* because (1) it can be easily extended to implement new data access methods; (2) it is capable of simulating different application environments; (3) new evaluation criteria can be added. The components of the testbed are described as follows:

Simulator: The *Simulator* object acts as the coordinator of the whole simulation process. It reads and processes user input, initializes data source, and starts broadcasting and request generation processes. It also determines which data access method to use according to the user input.

BroadcastServer: It is a process to broadcast data continuously. The *Broadcast-Server* constructs broadcast channel at the initialization stage according to the input parameters and then starts the broadcast procedure.

RequestGenerator: *RequestGenerator* is another process initialized and started by the *Simulator*. It generates requests periodically based on certain distribution. In the simulations covered by this proposal, the request generation process follows exponential distribution.

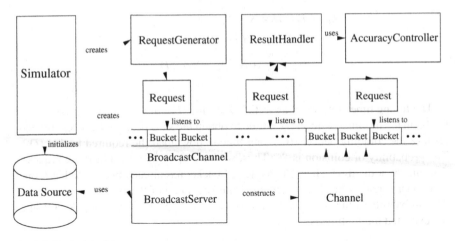

Fig. 3.5 Testbed Architecture

Data: Information to be broadcast is represented by a *Data* object. The infor-
             mation is read from files or databases. The *Data* object consists of a number of
             records.
Record: It is used to represent a broadcast data item. Each record has a primary
             key and a few attributes.
Bucket: Broadcast data items are reorganized as buckets to put in broadcast chan-
             nel. A bucket can be either index (signature) or data. The *Bucket* object is used
             to represent such bucket. Different access methods have different bucket organi-
             zation.
Channel: The *Channel* object consists of a number of *Bucket*s. It is constructed at
             initialization stage based on which data access method is being used. The *Broad-
             castServer* process broadcasts information contained in the broadcast *Channel*.
Request: User request is represented by *Request* objects. Each request is consid-
             ered to be an independent process. Once generated by the *RequestGenerator*, it
             starts listening to the broadcast channel for the required data item till it is found.
ResultHandler: In our testbed, we use access time and tuning time as criteria to
             evaluate the performance of the access methods. The *ResultHandler* object ex-
             tracts and processes the simulation results and output them in the proper format.
AccuracyController: To ensure the accuracy of our simulation, we use an *Accu-
             racyController* object to control the accuracy of the simulation results. Users can
             specify the accuracy expectation for the simulation. The simulation process will
             not terminate unless the expected accuracy is achieved.

We also implemented the analytical evaluation model of each access method in
the testbed so that we can compare the simulation results against the theoretical
results. The testbed supports the following environmental parameters that may have
influence on the performance of access methods:

N_r	Number of broadcast data items
S_{dk}	Key size of data items
S_d	Data item size
B_d	Broadcast channel bandwidth
D_a	Data availability
C_l	Confidence level
C_a	Confidence accuracy
D_r	Distribution for request interval

Table 3.2 Supported testbed parameters

The *data availability* defines the possibility of the requested information to be present in broadcast channel. In real settings, the requested information may not exist at all in the broadcast channel. How data availability affects performance becomes an important issue and worthwhile investigating. In applications with very low data availability, access methods requiring one-at-a-time browsing have usually very poor efficiency because clients scan the whole broadcast channel. Whereas for some other access method, such as distributed indexing, the performance is better with very low data availability because clients only need to scan the indexing section to determine the presence of requested information.

The *confidence level* and *confidence accuracy* [1] are used to control the accuracy of the simulation results. Users can specify the values of confidence level and accuracy before starting simulation. The simulation is not complete until the expected confidence level and accuracy are achieved.

The *request interval* parameter determines which distribution (e.g. exponential distribution) the testbed follows to generate requests. This parameter can be used to simulate different access patterns.

Figure 3.5 shows the generic architecture of the testbed. When a particular data access method is used, specific objects are created. For example, if user chooses to use signature indexing, the *BroadcastServer* constructs a *SigChannel* which would consist of a number of *SigBuckets*. The *RequestGenerator* would periodically generate *SigRequests*. The *Simulator* would create a *SigResultHandler* object to process the signature indexing specific results. The generic objects only store information that is common to all data access methods.

The testbed is implemented as a discrete event driven simulation system. The following procedures explain how the testbed works:

- Initialization stage:

  - Create *Simulator* object

---

[1] Given N sample results $Y_1, Y_2, ..., Y_N$, the *confidence accuracy* is defined as H/Y, where H is the *confidence interval half-width* and Y is the sample mean of the results ($Y = (Y_1 + Y_2 + ... + Y_N)/N$). The *confidence level* is defined as the probability that the absolute value of the difference between the Y and μ (the true mean of the sample results) is equal to or less than H. H is defined by $H = t_{\alpha/2;N-1} \times \sigma/\sqrt{N}$ where σ^2 is the sample variance given by $\sigma^2 = \Sigma_i (Y_i - Y)^2/(N-1)$ (thus σ is the standard deviation), and t is the standard t distribution.

- *Simulator* initializes the *Data* object, *BroadcastServer* object, *RequestGenerator* object and specific *ResultHandler* object.
- *Data* object reads the data source and creates all *Record* objects.
- Depending on which access method is selected, *BroadcastServer* creates the corresponding *Channel* object.
- Specific *Channel* object is created. A number of corresponding *Bucket* objects are created based on data records. Buckets are inserted into the broadcast channel according to the data organization of the chosen data access method.

- Start stage:

  - *Simulator* starts *BroadcastServer*.
  - *Simulator* starts *RequestGenerator*.

- Simulation stage:

  - Broadcast channel broadcasts the information continuously.
  - *RequestGenerator* generates *Request* periodically using the exponential distribution.
  - Generated requests listen to the broadcast channel and query the required information.
  - When completed, *Request* objects inform *ResultHandler* object of the results.
  - *Simulator* checks if the results are within the expected confidence and accuracy level to determine whether to terminate or continue the simulation.

- End stage:

  - *Simulator* stops *BroadcastServer*.
  - *Simulator* stops *RequestGenerator*.
  - *ResultHandler* processes the results and outputs them in a proper format.

A integrated GUI is developed for easy usage. Figure 3.6 shows a running example of the testbed GUI.

The graphical user interface (GUI) allows users to input parameters, execute the logic, and view the results. It also allows users the ability to display graphical plots (e.g. Cartesian graphs). The GUI has user controls for data storage, such as, save test results (input and output values), retrieve test results from previous test, and delete test results.

3.4 Practical Study

We now present some experiment results for a few existing wireless data access methods. We show both practical and theoretical results for each method. The practical results are obtained by running experiments using each access method, and the theoretical results are produced from the formulas presented in Section 3.2. By comparing practical results against theoretical results, we show the correctness of

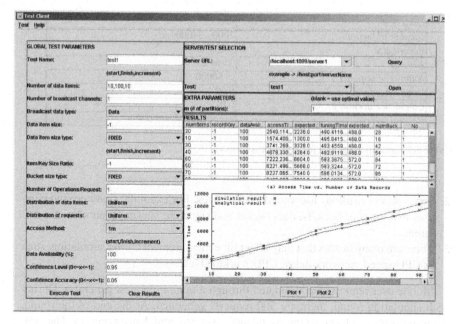

Fig. 3.6 Testbed Client GUI

our system implementation and the predictiveness of the analytical cost model. As a benchmark, we also show the experiment results for broadcasting without using any data access method, i.e., flat or plain broadcast.

3.4.1 Simulation Settings

The data source our testbed uses is a dictionary database consisting of about 35,000 records. Table 3.3 shows the simulation settings for all experiments presented in this section:

The testbed reads the data records from the data source and constructs them into data buckets using specific access methods. The access time and tuning time of each access method are evaluated in terms of the number of bytes read. Often possible parameters like the number of buckets or actual time elapsed are not used for the following reasons:

1. Bucket sizes may vary depending on the access method.

Data type	text
Number of records	7000 - 34000
Record size	500 bytes
Key size	25 bytes
Number of requests	> 50000
Confidence level	0.99
Confidence accuracy	0.01
Request interval	exponential distribution

Table 3.3 Simulation settings for data access methods

2. Some access methods use varied bucket sizes. For example, signature indexing uses two types of buckets, data bucket and signature bucket. Their sizes are different.
3. There are many factors that may affect time measurement during simulation, such as CPU speed, network delay, CPU workload, etc.

Mobile user requests are simulated through a random request generator. During each simulation round, there are 500 requests generated. At the end of the round, the result is checked against the confidence level and confidence accuracy. The simulation continues if the confidence conditions are not satisfied. In most of our simulation experiments, more than 100 simulation rounds are required to satisfy the confidence conditions we defined in Table 3.3. The generation of requests follows the exponential distribution. In all simulation experiments discussed in this section, we assume all requested data records are found in the broadcast.

3.4.2 Simulation Results

Based on the above simulation settings, each access method is simulated using the testbed. Figure 3.7 shows the simulation results for flat broadcast, distributed indexing, signature indexing and simple hashing. The lines marked with (S) are simulation results. Those marked with (A) are analytical results. We observe in both figures that the simulation results match the analytical results very well.

Flat broadcast exhibits the best access time but worst tuning time. With the introduction of data access methods, extra information is inserted into broadcast channel to help clients get to the required data item. This also introduces overhead in the broadcast cycle that consequently increases access time. In the flat broadcast, information is broadcast over the wireless communication channel without using any access method. Mobile clients must traverse all buckets to find the requested data. In such a case, mobile clients keep listening to the broadcast channel. This results in the worst tuning time. The expected average access time and tuning time are the same, which is approximately half of the broadcast cycle. Both access time and tuning time increase linearly with number of broadcast data records.

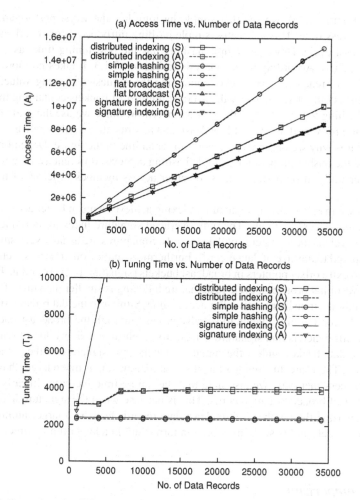

Fig. 3.7 Comparison for different access methods

From the formula presented in Section 3.2.2, since D_t is a constant, it is obvious that the tuning time of distributed indexing is determined by k, the number of levels in the index tree. The increase in the number of broadcast data records does not necessarily increase the tuning time, unless the bottom level of the index tree is filled and the number of levels in the index tree is, as a result, increased. The simulation result shows the same trend. We can also observe that the tuning time of distributed indexing only increases at one point (somewhere between 5000 and 10000 data records), where the number of levels in index tree is increased. The value of tuning time is much less than that of flat broadcast. With the help of indices, clients selectively listen to the broadcast channel, thus reducing the tuning time drastically.

From Figure 3.7(a), we see that simple hashing has the worst performance in terms of access time. This is because simple hashing introduces more overheads to the broadcast cycle. However, simple hashing has the best tuning time, as shown in Figure 3.7(b). According to the analysis in Section 3.1.3, it takes no more than four probes to reach the first bucket containing the requested hashing value. The tuning time is only determined by the average number of buckets containing the requested hashing value, which is the average overflow. Since we use the same hashing function for varied number of broadcast data records, the overflow rate is a fixed value. That is why we observe a straight horizontal line in the figure. Depending on how good the hashing function is, simple hashing achieves different average tuning times. Generally, it outperforms most of other access methods in terms of tuning time.

As shown in Figure 3.7(a), signature indexing achieves a much better access time than all other access methods. The only overhead signature indexing introduces in broadcast cycle is the signatures of data records. Signatures are usually very small in size compared to the size of data records. Furthermore, unlike other access methods, there is no extra offset pointers in broadcast buckets (signature buckets or data buckets). The resulting broadcast cycle of signature indexing is smaller than that of other access methods, translating in smaller access time. Similar to the flat broadcast, signature indexing requires clients to serially browse through the broadcast buckets. The only difference is that signature is read first each time. Clients do not read the successive data buckets unless the signature matches the signature of the requested data record. Therefore, the tuning time of signature indexing is much larger than that of other access methods. We also note that the tuning time increases linearly with the number of broadcast data records. This is because when the signature length is fixed, increasing the number of broadcast data records leads to a larger number of signature buckets and false drops, which in turn result in a larger tuning time.

3.4.3 Comparison

In this section, we compare the access methods under different scenarios. We define the following two parameters: (1) *Data availability* which defines the possibility of the requested information to be present in broadcast channel, and *Record/key ratio* which is the proportion of record size to key size.

In real settings, the requested information may not exist at all in the broadcast channel. How data availability affects performance becomes an important issue. In applications with very low data availability, access methods requiring one-at-a-time browsing have usually very poor efficiency because clients scan the whole broadcast channel. In other access methods, such as distributed indexing, however, clients only need to scan the indexing section to determine the presence of requested information.

The record/key ratio is another important factor that may affect the efficiency of different access methods. For B+ tree based access methods, such as distributed indexing, higher record/key ratio implies more indices likely to be placed in a single bucket, which in turn would reduce the number of index tree levels. As previously discussed, tuning time is mainly determined by the number of index tree levels in these two access methods. Therefore, record/key ratio has a substantial influence on the performance of B+ tree based access methods. Record/key ratio has also great impact on the efficiency of signature or hashing based access methods. Smaller record/key ratio usually means less overhead being introduced to the broadcast cycle, which results in better access time. However, smaller record/key ratio may also lead to more false drops in signature indexing and higher overflow in hashing, causing worse tuning time. In real world applications, record/key ratios in different wireless applications may vary largely. Therefore, it is important to study how record/key ratio affects the efficiency of access methods.

3.4.3.1 Comparing access methods Based on Data Availability

We vary data availability from 0% to 100% in our simulation to evaluate the performance of different access methods. Figure 3.8 shows the result of access time and tuning time against data availability.

Figure 3.8(a) clearly shows that data availability in hashing has little impact on access time. This is because changing the data availability does not change broadcast cycle and the access sequence of hashing method is only determined by the hashing function. We note that when the data availability is high (towards 100%), flat broadcast and signature indexing have the best performance. When the data availability is low (towards 0%), *(1,m)* indexing and distributed indexing outperform all other methods. The reason is that *(1,m)* indexing and distributed indexing can determine the presence of the requested information by reading only the index segment.

As to tuning time, we do not consider it for flat broadcast simply because it is much larger than that of all other methods. Tuning time of signature indexing decreases with the increased data availability. This is because when data availability increases, there is less probability that clients scan the whole broadcast cycle to find out whether the requested record is in the broadcast. Figure 3.8(b) shows that *(1,m)* and distributed indexing perform better under low data availability, whereas signature indexing and hashing have better performance when data availability is high. For *(1,m)* and distributed indexing, it only takes a few probes (by reading only a few levels of index tree) to determine whether the requested information is present. Therefore, they have the best performance when there is very low data availability. However, when data availability is high, extra probes are needed to find the actual data buckets in the broadcast. Therefore, the tuning time increases with data availability. For the hashing method, all overflow buckets must still be read when the requested information is not in the broadcast. It outperforms *(1,m)* indexing and distributed indexing at high data availability because it may need to read fewer overflow buckets before reaching the required bucket.

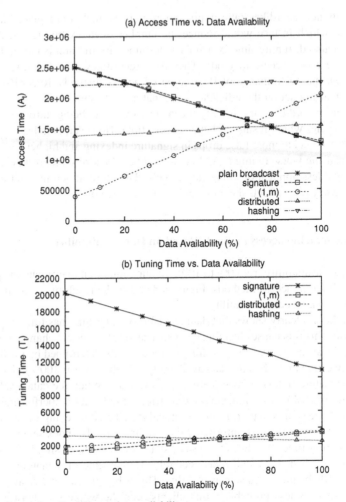

Fig. 3.8 Comparison for different data availability

3.4.3.2 Comparing access methods Based on Record/Key Ratio

In this experiment, we assume that data availability is achieved at 100%. We vary the record to index ratio from 5 to 100. This is a reasonable range that record/key ratios of data in most applications fall in. The simulation results are shown in Figure 3.9.

Figure 3.9(a) shows that access time changes with the record to key ratio. We see that the ratio has a strong impact only on *(1,m)* indexing and distributed indexing. That is because different record to key ratios may result in different index tree structures. For B+ tree based access methods, the ratio changes the number of buckets in the index tree and thus the index levels. For distributed indexing, the ratio also changes the number of replication levels. Figure 3.9(a) shows that both *(1,m)* indexing and distributed indexing have very large access times when the record to key ratio is small. This is because the overhead introduced by adding the index segment, is prominent when the key size is comparable with the record size. They perform much better when the ratio is large. Both *(1,m)* indexing and distributed indexing outperform hashing as record/key ratio gets larger.

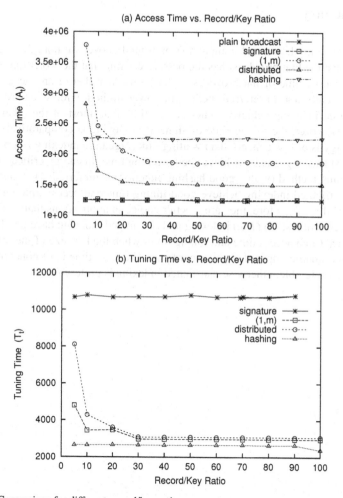

Fig. 3.9 Comparison for different record/key ratio

As to tuning time, flat broadcast is ignored because it is not comparable with the other methods. From our previous discussion, the tuning time of hashing is much better than that of other access methods. Figure 3.9(b) shows that *(1,m)* indexing and distributed indexing exhibit similar performance to hashing when the record/key ratio is large. The explanation is similar to that of access time. When the key size is large compared to the data record, fewer indices fit into a single bucket. Thus, more buckets are read to traverse the index tree. This results in a larger tuning time. On the other hand, smaller key sizes result in smaller tuning time for *(1,m)* indexing and distributed indexing.

3.5 Summary

Based on the simulation and comparison presented above, we note the following observations: (1) Flat broadcast has the best access time but the worst tuning time. Since the tuning time of flat broadcast is far larger than that of any access methods, it is usually not a preferred method in power limited wireless environments; (2) Signature indexing achieves better access time than most of the other access methods. However, the tuning time of signature indexing is comparatively larger. When energy is of less concern than waiting time, signature indexing is a preferred method; (3) Hashing usually achieves better tuning time. In energy critical applications, hashing method (with a good hashing function) is preferred; (4) *(1,m)* indexing and distributed indexing achieve good tuning time and access time under low data availability. Therefore, they are a better choice in applications that exhibit frequent search failures; (5) *(1,m)* indexing and distributed indexing have good overall performance (both access time and tuning time) when the key size of the broadcast data is very small compared to the record size. If access time is of concern, *(1,m)* indexing is preferable, otherwise distributed indexing is preferable.

Chapter 4
Adaptive Data Access Methods

In this chapter, we present an adaptive method we proposed to improve efficiency of wireless data access. The proposed method is based on the observation that the index tree based methods preserve good access time as well as stable overall performance and that the hashing method exhibits good tuning time. By combining these two techniques, the new method takes the advantages of both techniques. As a result, it exhibits greater flexibility and better performance.

As previously shown, the tuning time of hashing depends on N_c, which in turn depends on the number of collisions. The number of collisions normally depends on how good the hashing function is. Thus, deriving a good hashing function is crucial for good tuning time. However, because of the heterogeneity of broadcast data in different applications, this is usually difficult to achieve. Furthermore, the hashing function itself is included in every data bucket. This obviously increases the broadcast cycle, and thus, the access time. In our method, hashing is used only to partition the broadcast data into a number of partitions. B+ tree technique is then used to index each partition. The hashing function is only stored at the beginning of each partition. Since the number of partitions is a small number compared to the number of all data items, the overhead introduced is much smaller than that in hashing based method. We now show how hashing and index tree techniques are combined in our method:

- *First level hashing*: Generate hashing value (h_1) for key attribute of each data item using a hashing function H_1. The hashing value generated at this step is similar to the hash value in simple hashing method.
- *Second level hashing*: Use another hashing function (H_2) to produce p second level hashing values (h_2) based on the value of h_1 generated in the first step. This process will partition the broadcast data into p parts. All data items having the same second level hashing value will be in the same partition.
- *Generating the index tree*: Within each partition, generate an index tree on the key attributes of the data items in the partition.

X. Yang and A. Bouguettaya, *Access to Mobile Services*, Advances in Database Systems 38, DOI: 10.1007/978-0-387-88755-5_4, © Springer Science + Business Media, LLC 2009

The two level hashing can actually be combined into one hashing method. We split it into two levels to better illustrate the steps. The adaptive method has the following advantages compared with simple hashing and distributed indexing:

- Hashing is used only to partition the broadcast data. The number of partitions is a small number compared to the size of data items. Thus, it is much easier to find hashing functions that generate hash codes with low conflicts on partitions than on all data items.
- Hashing functions need only be included in front of each data partition. This reduces the broadcast cycle and as a result, the access time.
- The index tree levels of data items in one partition is smaller than that of all data items (the exact number is determined by p). This may result in an improvement in tuning time because the tuning time is determined by the levels of index tree in each partition plus the overhead introduced by reading the hashing functions.

4.1 Data Organization

In the adaptive method, a data partition together with its associated index tree form a *broadcast partition*. At the beginning of each broadcast partition, hashing functions and the hashing value of this partition are stored. This information is used to locate the partition that contains the requested data item. Figure 4.1 illustrates how the broadcast channel is organized for the adaptive method.

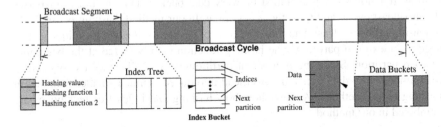

Fig. 4.1 Channel structure of adaptive method

Hashing is used to locate the partition that the requested data item belongs to. The index tree helps to find the data item in the partition. Users only need to check the hashing value at the beginning of each partition to determine if the requested data item is in the coming partition. A match implies the requested data item is in this partition. If the hashing values do not match, the request will be forwarded to the right partition.

The index tree of each partition is placed right after the hashing bucket(s) in the broadcast channel. This is different from distributed indexing where the index tree is partitioned. Instead, the whole tree (only for the data items in this partition) is used because there is no sub-partitions. Data buckets are placed following the index tree in each partition. Each bucket, index or data, has a pointer entry pointing to the next nearest partition.

To divide the broadcast cycle into p partitions using hashing, we first use one or more hashing functions to generate hashing values H ($1 \leq H \leq p$) for each data item. We then group every data item with the same hashing value into one partition. The requirement for choosing the hashing functions is the ability to generate partitions containing the same approximate size of data items across partitions. We use a two level hashing method to achieve this objective. First, each key field is converted into its integer representation using the ASCII standard. The resulted number is then hashed to the range $[1, p]$. The optimum scenario is when a hashing function produces partitions with equal size. In this case, we only need to store a hashing value and a hashing function at the beginning of each partition. The required size in this case is much smaller than a bucket. It is usually difficult to achieve this goal in real settings. We propose two variants of the adaptive method for non-optimum scenarios: (1) *Fixed Partition Size* (FPS) adaptive method, and (2) *Varied Partition Size* (VPS) adaptive method.

Fixed Partition Size Adaptive Method

As the name of this method implies, the size of all partitions is the same. Since the hashing functions do not necessarily generate the same number of data items in each partition, empty buckets are appended to the partitions to make them the same size as the largest partition. As a result, the only hashing information that needs to be stored at the beginning of each partition is the partition hashing value and the hashing functions. Since all partitions are of the same size, the location of each partition is computed by the hashing functions. The empty buckets obviously introduce extra overhead into the broadcast cycle. However, this overhead should be low because of the large data size.

Varied Partition Size Adaptive Method

In this method, the partitions are of different sizes. To locate a partition, a mapping directory is stored at the beginning of each partition. This mapping directory maps each hashing value to the location of its corresponding partition. The hashing value of the requested data item is calculated first by applying the stored hashing functions. The hashing value is then looked up in the mapping directory to obtain the location of the corresponding partition. When compared with the FPS method, the VPS method eliminates the overhead introduced by the empty buckets. However, it introduces an overhead in the broadcast cycle by storing the mapping directory at the beginning of each partition. This method is preferred when the partition sizes have a large variance.

4.2 Access Protocol

We now present the access protocols of both FPS and VPS methods. We use the request for a data item with key K as a sample scenario.

Access Protocol for the FPS Method

```
        Mobile client requests for data with key K
        Tune into the broadcast channel
        Keep listening until the first complete bucket
            arrives
        Read the bucket
        Go to the next partition according to the offset
            value in the first bucket
        Read the hash bucket and get the hashing functions
            H1 and H2.
    (1) Read the hashing value h of the current partition
        If h < H2(H1(K))
          Go to doze mode
          Tune in again when h = H2(H1(K)) (the partition
              with the right hashing value)
        Else
          Go to doze mode
          Tune in again at the beginning of the next
              broadcast
          Repeat from (1)
    (2) Read the bucket in the index tree
        If K = Key being broadcast
          Read time offset to the data bucket
          Go into doze mode
          Tune in again at the data bucket
          Download the data bucket
        Else
          Read index in current index bucket
          Follow the index path to go to the lower level
              index bucket
          Go into doze mode between any two successive
              index probes
```

```
        Repeat from (2)
```

Access Protocol for the VPS Method

The access protocol of the VPS method is mostly similar to the FPS method except
on how the correct partition is found.

```
        Mobile client requests data with key K
        Tune into the broadcast channel
        Keep listening until the first complete bucket
            arrives
        Read the bucket
        Go to the next partition according to the offset
            value in the first bucket
        Read the hash bucket and get the hashing functions
            H1 and H2.
    (1) Read the hashing value h of the current partition
        If h < H2(H1(K))
            Read the location of each partition in current
                bucket and all hash buckets that follows
                till the location of the partition with
                hashing value H2(H1(K)) is found.
            Save the offset value to that partition
            Go to doze mode
            Tune in again when the partition arrives
        Else
            Go to doze mode
            Tune in again at the beginning of the next
                broadcast
            Repeat from (1)
    (2) Read the bucket in the index tree
        If K = Key being broadcast
            Read the time offset to the data bucket
            Go into doze mode
            Tune in again at the data bucket
            Download the data bucket
        Else
            Read index in current index bucket
            Go to lower level index bucket following the
                index path
            Go into doze mode between any two successive
                index probes
            Repeat from (2)
```

4.3 Cost Model

In this section, we provide an analytical cost model for the proposed adaptive methods. Table 4.1 shows the symbols definitions used in the cost model.

p	Number of partitions
N_r	Total number (size) of data items to be broadcast
N	Total number of buckets in a broadcast cycle
S	Bucket size
$N(j)$	Number of buckets in partition j
$N_i(j)$	Number of index buckets in partition j
$N_r(j)$	Number of data buckets in partition j
$N_e(j)$	Number of empty buckets in partition j
N_h	Number of buckets containing hashing information in each partition
$k_p(j)$	Index tree levels for partition j
n_p	Number of index entries in an index bucket
n_h	Number of mapping entries in a hashing bucket

Table 4.1 Symbols for adaptive access methods

4.3.1 Derivation for Access and Tuning Times

Each partition consists of data buckets, index buckets, and buckets containing hashing information (i.e. hashing buckets). For the FPS method, the hashing buckets are used to store the hashing value of the partition and the hashing functions. For the VPS method, the hashing buckets also store the mapping directory that helps users to find the location of a partition. For the FPS method, empty buckets are inserted into every partition (except the largest one) to make all partitions equal in size. The, the number of buckets in each partition can be expressed as:

$$N(j) = N_i(j) + N_r(j) + N_e(j) + N_h$$

where $N_e(j) = 0$ for the VPS method. Let $k_p(j)$ be the levels of index tree in each partition, and n_p the number of indices in an index bucket. We have

$$k_p(j) = \lceil \log_{n_p} N_r(j) \rceil$$

The number of index buckets (N_i) in each partition can be calculated as:

$$N_i(j) = 1 + n_p^2 + \ldots + n_p^{k_p(j)-1} = \frac{n_p^{k_p(j)} - 1}{n_p - 1}$$

As discussed above for the FPS method, there are empty buckets in the partitions to make the size of every partition equal. The length of each partition, $N(j)$, for the FPS method is:

$$N(j) = N_r(j) + N_i(j) + N_e(j) + N_h$$
$$= N_r(j) + \frac{n_p^{\lceil \log_{n_p} N_r(j) \rceil} - 1}{n_p - 1} + N_e(j) + N_h \qquad (4.1)$$

For the FPS method, N_h is normally equal to 1. $N(j)$ for the VPS method can also be expressed using the same formula except that $N_e(j) = 0$, and N_h usually has larger values (because of the mapping directory).

The access times of both fixed and varied adaptive methods include the following parts:

1. *Initial wait (F_t)*: This part is common to all access methods. It is the time for the first complete bucket to arrive, which is on average $\frac{D_t}{2}$.
2. *Initial partition probe (IPP_t)*: This is the time spent to reach the next nearest partition. The average time should be the time to retrieve half of a partition, which can be expressed as:

$$IPP_t = \frac{1}{2p} \times D_t \times \sum_{j=1}^{p} N(j)$$

3. *Partition probe (PP_t)*: The partition probe is the time period from the first partition arrival to the partition containing the requested data item. The average partition probe is half of the whole broadcast cycle, which is:

$$PP_t = \frac{1}{2} \times D_t \times \sum_{j=1}^{p} N(j)$$

4. *Data probe (DP_t)*: The data probe is the time to find the requested data item within a partition. The average data probe is the time to retrieve half of a partition, which can be expressed using the same formula as IPP_t.

Therefore, the access time can be calculated as:

$$A_t = F_t + IPP_t + PP_t + DP_t$$
$$= \frac{D_t}{2} + \frac{1}{2p} \times D_t \times \sum_{j=1}^{p} N(j) + \frac{1}{2} \times D_t \times \sum_{j=1}^{p} N(j)$$
$$+ \frac{1}{2p} \times D_t \times \sum_{j=1}^{p} N(j)$$
$$= (\frac{1}{2} + (\frac{1}{p} + \frac{1}{2}) \times \sum_{j=1}^{p} N(j)) \times D_t \qquad (4.2)$$

The tuning time can be calculated similarly. It includes: the initial wait ($F_t = \frac{D_t}{2}$), one probe to reach the first partition (D_t), maximum N_h probes to obtain the hashing value to reach the right partition ($N_h \times D_t$), maximum $k_p(j)$ probes in index tree to obtain the location of the data item ($k_p(j) \times D_t$), and one more probe to download the actual data item (D_t). Therefore, the tuning time is:

$$
\begin{aligned}
T_t &= \frac{D_t}{2} + D_t + N_h \times D_t + k_p(j) \times D_t + D_t \\
&= (k_p(j) + N_h + \frac{5}{2}) \times D_t
\end{aligned}
\tag{4.3}
$$

4.3.2 Optimum Number of Partitions

The behavior of the proposed methods relies on the number of partitions, p. When p is small, the behavior of the adaptive methods is similar to that of the index tree based methods. However, the best access time is not achieved when $p = 1$. This is because when p increases, the initial partition probe (IPPt) time decreases. The initial partition probe is the time spent to reach the next nearest partition. The average time should be the time to traverse half of a partition. As p gets larger, each partition size gets smaller. Thus, the initial partition probe time becomes smaller. This in turn causes the overall access time to decrease. In contrast, as p gets larger, the overall broadcast cycle also gets longer. For FPS, this is due to the addition of more empty buckets. For VPS, this is due to the mapping directory getting larger because more buckets are now stored in each partition. When the increase of the broadcast cycle starts to have a greater impact than the decrease of the initial partition probe, the access time would start to increase as p increases. When p is large (i.e., the broadcast data is partitioned into fine granules), the behavior of the proposed methods is similar to that of the hashing method. For the FPS method, the tuning time is better as p increases. In contrast, the tuning time for the VPS method is affected by the length of the mapping directory at the beginning of each partition. Note that this length increases with p. This has a potential negative effect on the tuning time.

In what follows, we study the impact of the number of partitions, p has on (1) the access time of the FPS method, and (2) the access and tuning times of the VPS method.

We use $p_f(AT)^*$ to represent the value of p when the access time for the FPS method is minimum. We use $p_v(AT)^*$ and $p_v(TT)^*$ to represent the value of p when the VPS method has the best access time and tuning time respectively. Let us first derive the value of $p_f(AT)^*$. In Formula (4.2), we can intuitively see that $\sum_{j=1}^{p} N_r(j) = N_r$ and $n_p^{\lceil \log_{n_p} N_r(j) \rceil} \approx N_r(j)$. For the FPS method, the total number of empty buckets in a broadcast cycle is $\sum_{j=1}^{p} N_e(j)$ This is not dependent on p, but rather on the hashing function itself. Therefore, we have:

$$\sum_{j=1}^{p} N(j) = \sum_{j=1}^{p} (N_r(j) + \frac{n_p^{\lceil \log_{n_p} N_r(j) \rceil} - 1}{n_p - 1} + N_e(j) + N_h)$$

$$= N_r + \frac{N_r - p}{n_p - 1} + N_e + N_h \times p \qquad (4.4)$$

Combining (4.2) and (4.4), we obtain:

$$A_t = (\frac{1}{2} + (\frac{1}{p} + \frac{1}{2}) \times (N_r + \frac{N_r - p}{n_p - 1} + N_e + N_h \times p)) \times D_t$$

$$= (\frac{1}{p} \times (N_e + \frac{n_p \times N_r}{n_p - 1}) + \frac{p}{2} \times (N_h - \frac{1}{n_p - 1}) + C_1) \times D_t \qquad (4.5)$$

where $C_1 = \frac{1}{2} \times N_e + \frac{n_p \times N_r - 2}{2 \times (n_p - 1)} + N_h$. The best value of A_t is achieved when the derivative of Formula (4.5) on p is 0. The value of $p_f(AT)^*$ can be calculated as:

$$p_f(AT)^* = \sqrt{\frac{2 \times (N_e + \frac{n_p \times N_r}{n_p - 1})}{N_h - \frac{1}{n_p - 1}}} \qquad (N_h \geq 1, n_p > 2) \qquad (4.6)$$

The value of $p_v(AT)^*$ can be derived similarly from Formula (4.2). Let n_h be the number of mapping entries a bucket can store. The number of buckets used to store the mapping directory in each partition would be $\lceil \frac{p}{n_h} \rceil$. For the VPS method, $N_e(j) = 0$. Thus, we have:

$$\sum_{j=1}^{p} N(j) = \sum_{j=1}^{p} (N_r(j) + \frac{n_p^{\lceil \log_{n_p} N_r(j) \rceil} - 1}{n_p - 1} + \lceil \frac{p}{n_h} \rceil)$$

$$= N_r + \frac{N_r - p}{n_p - 1} + \frac{p^2}{n_h} \qquad (4.7)$$

Combining (4.2) and (4.7), we obtain:

$$A_t = (\frac{1}{2} + (\frac{1}{p} + \frac{1}{2}) \times (N_r + \frac{N_r - p}{n_p - 1} + \frac{p^2}{n_h})) \times D_t$$

$$= (\frac{1}{2 \times n_h} \times p^2 + (\frac{1}{n_h} - \frac{1}{2 \times (n_p - 1)}) \times p + \frac{n_p \times N_r}{n_p - 1} \times \frac{1}{p} + C_2) \times D_t \quad (4.8)$$

where $C_2 = \frac{1}{2} \times (1 + N_r + \frac{N_r - 2}{n_p - 1})$. The best access time is achieved when the derivative of Formula (4.8) is 0, which can be expressed as:

$$(\frac{p}{n_h} + (\frac{1}{n_h} - \frac{1}{2 \times (n_p - 1)}) - \frac{n_p \times N_r}{n_p - 1} \times \frac{1}{p^2}) = 0$$

i.e. $$\frac{1}{n_h} \times p^3 + (\frac{1}{n_h} - \frac{1}{2 \times (n_p - 1)}) \times p^2 - \frac{n_p \times N_r}{n_p - 1} = 0 \qquad (4.9)$$

The best access time of the VPS method is obtained by solving Formula (4.9):

$$p_v(AT)^* \approx \sqrt[3]{n_h \times N_r} \qquad (4.10)$$

We now derive the value of p to achieve the best tuning time performance for the VPS method. As stated above, the number of hashing buckets in each partition is $\lceil \frac{p}{n_h} \rceil$. On average, the wait is equivalent to traversing half of the hashing buckets to find the partition that contains the requested data item. This is $\frac{1}{2} \times \lceil \frac{p}{n_h} \rceil$. Replacing this value in Formula (4.3), we obtain:

$$T_t \approx (\log_{n_p} \frac{N_r}{p} + \frac{p}{2 \times n_p} + \frac{5}{2}) \times D_t \qquad (4.11)$$

Finally, the derivative of Formula (4.11) is:

$$p_v(TT)^* = \frac{2 \times n_p}{\ln n_p} \qquad (4.12)$$

4.4 Practical Study

In this section, we compare the proposed adaptive methods with hashing and index tree based tree methods using simulation. We first introduce the simulation settings we use for the comparison. We then present the results of the comparison. Finally, we propose a new approach to compare the overall performance of data access methods. We show that the adaptive methods have the best overall performance. Table 4.2 shows the common simulation settings used for all experiments presented in this section.

Data source	text database
Data type	text
Request distribution	uniform & nonuniform
Number of records	1,000 - 100,000
Record size	500 bytes
Key size	25 bytes
Hash value size	4 bytes
Pointer size	4 bytes
Query type	Non-range queries
Data availability	100%
Number of requests	$> 50,000$
Confidence level	0.99
Confidence accuracy	0.01
Request interval	exponential distribution

Table 4.2 Simulation settings for adaptive access methods

4.4.1 Access and Tuning Times vs. the Number of Partitions

In this section, we study how access and tuning times of the proposed methods vary as a function of the number of partitions (p). Figure 4.2 (a) and (b) shows the simulation results of access and tuning times against p for the proposed methods when the size of data items (N_r) is 50,000.

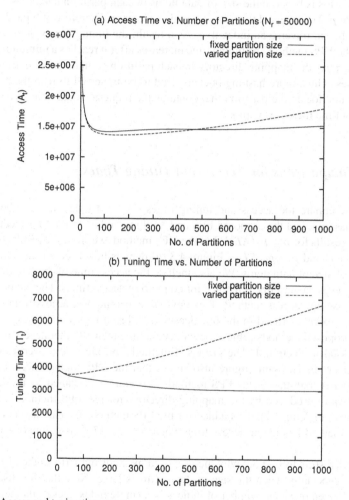

Fig. 4.2 Access and tuning times vs. p

As we can observe from Figure 4.2(a), the access time of both methods first decreases when p increases. After the access time reaches a minimum value, it starts to increase with p. This trend matches with our analysis in Section 4.3.2. We can also notice that the access time of the VPS method increases faster than that of the FPS method. This is due to the fact that there is a mapping directory in each partition for the VPS method. When p increases, both the number and the sizes of the mapping directories would increase in a broadcast cycle. Therefore, more buckets are needed in a broadcast cycle to store the mapping directories. This overhead causes the access time to climb faster.

As shown in Figure 4.2(b), the tuning time of the FPS method decreases when p increases. This is because the size of data items in each partition decreases with the increase of p. The levels of the index trees also decrease. Therefore, it typically takes fewer probes to traverse an index tree. As a result, the tuning time would decrease too. For the VPS method, the tuning time increases after it reaches a minimum value. This is because the mapping directory in each partition contains more entries when p increases. Thus, more hashing buckets need to be traversed to read the mapping directory and locate the partition that contains the requested data item. As a result, the tuning time increases with p.

4.4.2 Comparisons for Access and Tuning Times

We now compare the access and tuning times of (1,m) indexing, distributed indexing, hashing, and the proposed FPS and VPS methods. For the FPS method, we show the results for $p = p_f(AT)^*$. For the VPS method, we show the results for both $p = p_v(AT)^*$ and $p = p_v(TT)^*$. Figure 4.3 (a) depicts how access time varies as a function of size of data items. We also include the access time of the *flat broadcast* in the graph as the baseline reference for comparing access times. The flat broadcast does not use any access method. This obviously has the best access time because there is no extra overhead in the broadcast cycle. From Figure 4.3 (a), we can see that the proposed methods achieve better access time than all other access methods. As discussed in Section 4, The proposed methods introduce less overhead in the broadcast cycle. The same figure also shows that the VPS method access time is slightly better than that of the FPS method. This can be explained by the fact that the overhead introduced by the mapping directory for the VPS method is usually smaller than the empty buckets added for the FPS method. We can also notice that the VPS method has better access time when $p = p_v(AT)^*$ as compared to when $p = p_v(TT)^*$.

It is interesting to find out why the proposed methods outperform the distributed indexing, especially when the size of data items is large. Note that for distributed indexing, given the same number of indices in a single index bucket, the number of levels of the index tree increases with the size of data items (buckets). This typically would result in an increase in the number of replicated index levels. Thus, we should expect that there would be more replicated index buckets at the beginning of each data segments. This would imply more overhead in the broadcast cycle. Compared with the distributed indexing, the increase in the size of of data items does not affect the access time of the proposed methods as much. For the FPS method, the overhead

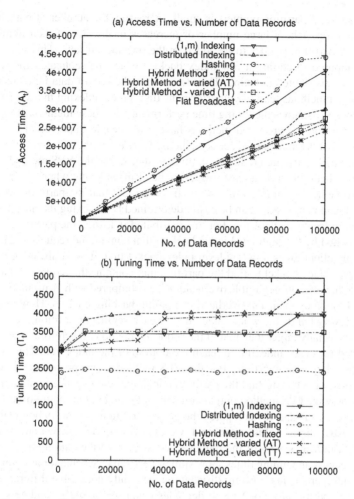

Fig. 4.3 Comparison of access and tuning times

comes from two factors: (1) hashing buckets at the beginning of each partition, and
(2) the empty buckets to make partition sizes equal. Note that the hashing buckets at
the beginning of each partition are used to store the hashing functions and hashing
value of the current partition. The number of hashing buckets is not affected by
the size of data items. The number of empty buckets may vary in each partition.
However, this is independent from the size of data items. Therefore, the increase in
the size of data items does not introduce extra overhead in the broadcast cycle for the
FPS method. For the VPS method, the size of data items does have an impact on the
number of hashing buckets. However, this impact is usually negligible, especially

when the size of partition pointers is small. Increasing the number of partitions by one, which may take a large number of increases in the size of data items, only causes one partition pointer to be added in the hashing buckets. This is relatively small compared to adding a few index buckets when the replicated index levels increase for distributed indexing. Therefore, when the size of data items increases, the proposed methods tend to *outperform* the distributed indexing in access time.

Figure 4.3 (b) shows the tuning time performance for these methods. The results demonstrate that the proposed methods have a noticeable improvement in tuning time over the index tree based access methods. The FPS method *clearly* outperforms both (1,m) and distributed indexing in tuning time. As for the VPS method, it has a tuning time better than or at least similar to the (1,m) indexing when $p = p_v(TT)^*$. The improvement in tuning time comes from the fact that the partition where the requested data item resides can be easily determined by checking the hashing buckets that is in front of a partition. For the distributed indexing, the partition can only be determined by the high level index buckets that contain the requested data item. If a mobile client tunes to a low level index bucket first, it must always go up to a higher level index bucket to find the partition that contains the requested data item. This introduces a non-negligible overhead. When compared with (1,m) indexing, the proposed methods incur the overhead of reading hashing buckets. However, since (1,m) indexing builds the whole index tree in each segment (partition), the number of levels is usually larger than that of the proposed methods. Therefore, more probes are needed for (1,m) indexing to reach the bottom of the index tree. This typically results in a larger tuning time. For the VPS method, extra probes are needed on the mapping directory to find the partition containing the requested data item. This explains how the FPS method outperforms the VPS method in tuning time.

Figures 4.4 (a) and (b) compares the access and tuning times when using Zipf request distribution. In this experiment, we use the empirical skew condition value 0.8, which indicates 80% of the requests access 20% of the data items. As can be seen, the access and tuning times show similar performance behavior as compared to their counterparts using uniform distribution. The only noticeable difference is the tuning time performance of VPS. It degrades when $p = p_v(AT)^*$, and gets slightly better when $p = p_v(TT)^*$.

Figure 4.5 (a) and (b) shows the impact of skew condition on the performance of access and tuning times for the proposed methods when using Zipf request distribution. In this experiment, we fix the number of data items and observe the performance changes while the skew condition varies from 0.0 to 1.0. As shown in the figures, the performances of both methods are stable under different skew conditions. This implies that the performances of the proposed methods are not affected by different access patterns of user requests.

We generally observe from Figures 4.3, 4.4, and 4.5 that under both uniform and non-uniform request distributions the proposed methods do take advantage of (1) hashing to reduce tuning time, and (2) index tree based techniques to reduce access time. The proposed methods also exhibit stable performance under different access patterns. The figures show that the proposed methods can achieve better access time

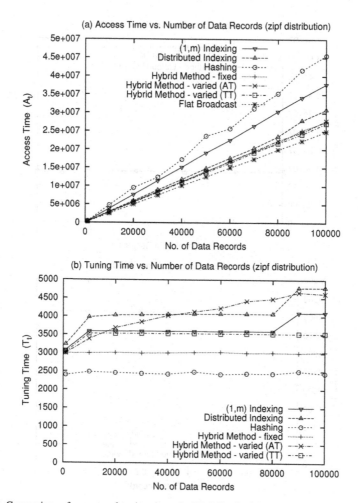

Fig. 4.4 Comparison of access and tuning times for Zipf distribution

than all other access methods. The proposed methods can also outperform the index tree based methods in tuning time.

4.4.3 Average Time Overhead Comparison

Access and tuning times are usually compared separately to reflect different aspects of wireless data access performance. In some applications, we might need to consider access and tuning times as a whole. This calls for a meaningful method to

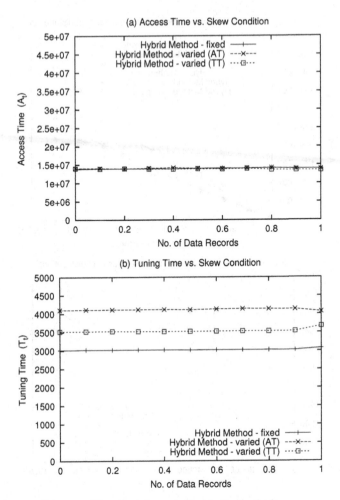

Fig. 4.5 Impact of skew condition on performance for adaptive methods

combine the results of access time and tuning time. Scanning time on 100 buckets, for example, may well possibly be ignored in access time for a large size of data items. However, it may make a huge difference in tuning time. Thus, directly using linear summation of access time and tuning time may not be meaningful. We propose a simple approach to compare access and tuning times for the different data access methods.

Approach: Given a list of access methods, assume their access and tuning times are $AT(i)$ and $TT(i)$, where i varies from 1 to n, representing n different data access methods. Let AT_{min} and TT_{min} be the best access and tuning times of all these methods, respectively. These two values may come from two different methods. We define **access time overhead** (AT_o) as the overhead of access time compared to the best access time value. We define **tuning time overhead** (TT_o) as the overhead of tuning time compared to the best tuning time value. Last, we define **average time overhead** (TO_{avg}) as the average overhead for both access and tuning times. For a data access method i, AT_o, TT_o, and TO_{avg} can be calculated as follow:

$$AT(i)_o = \frac{AT(i) - AT(min)}{AT(min)}$$

$$TT(i)_o = \frac{TT(i) - TT(min)}{TT(min)}$$

$$TO(i)_{avg} = \frac{1}{2} \times (AT(i)_o + TT(i)_o)$$

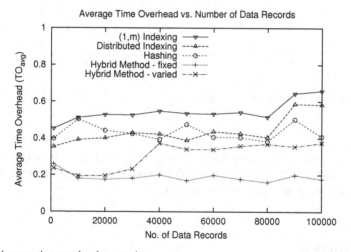

Fig. 4.6 Average time overhead comparison

Figure 4.6 shows the comparison of the average time overhead of the covered data access methods. We observe that the proposed methods have less average time overhead than all other methods. The proposed methods have the best average access time and tuning time performance.

In some applications, it may be important to know what the overall performance is if access and tuning times have differing importance. In this case, we assign different weights to access time and tuning time depending on the application requirements. The following derivation shows how to calculate average time overhead with weighted access and tuning times:

Derivation: *Given the weight α for access time and β for tuning time, the average overhead time for access method* i *is:*

$$TO(i)_{avg} = \alpha \times AT(i)_o + \beta \times TT(i)_o$$

where α and β are weights and $\alpha + \beta = 1$.

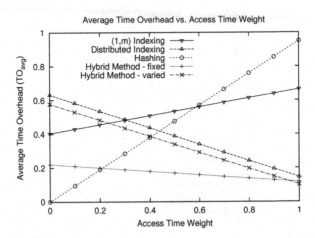

Fig. 4.7 Average time overhead with weight

Figure 4.7 shows the comparison of average time overhead with weighted access and tuning times at 50,000 data items. We observe that the proposed methods have the best performance under most weight values.

4.4.4 Summary

Based on the experiments presented above, we note the following observations: (1) The proposed adaptive access methods demonstrate better access time performance than hashing and index tree based methods; (2) The proposed adaptive access methods demonstrate better tuning time performance than the index tree based methods; (3) The VPS method shows better access time performance than the FPS method; (4) The FPS method shows better tuning time performance than the VPS method; (5) The access and tuning time performances of the proposed adaptive methods can be controlled by vary the number of partitions p; (6) The access and tuning time performances of the proposed adaptive methods are not affected by access frequencies; (7) In general, the proposed adaptive methods demonstrate better overall performance when considering both access and tuning times.

Chapter 5
Efficient Access to Simple M-services Using Traditional Methods

In this chapter, we discuss accessing simple M-services using traditional wireless access techniques. First, we show how different access methods can be applied to M-services environment. Then, we present analytical cost model for each access method.

As already mentioned, there are two important factors that are normally used to measure the performance of data access in wireless environments: *Access Time* and *Tuning Time*. Applying this concept to M-service environment, the access and tuning times reflect two important aspects of access to M-services, the response time and energy efficiency. In the context of M-services, we define access time and tuning time as follows:

- **Access Time:** This is the total client waiting time, starting from the issuance of a service request till its completion.
- **Tuning Time:** Tuning time is the time when CPU and/or wireless receiving devices are active, which means mobile clients are either actively processing requests and/or retrieving data from wireless channels. Tuning time includes the time to download or update the registry (if required), find and download the requested M-service, execute the M-service, and retrieve the requested data items.

5.1 Access Methods

We now show how traditional data access methods can be used to access simple M-service. For each data access method, we consider the following three scenarios: (1) access to the M-service channel with fixed-size M-services; (2) access to the M-service channel with varied-size M-services; and (3) access to the data channel. We make a reasonable assumption that data items in the data channel have the same size.

X. Yang and A. Bouguettaya, *Access to Mobile Services*, Advances in Database Systems 38, DOI: 10.1007/978-0-387-88755-5_5, © Springer Science + Business Media, LLC 2009

5.1.1 Index tree based methods

Here, we show how index tree based methods can be applied to M-service environment. We use (1,m) indexing instead of distributed indexing because of its simplicity.

Generally, we refer to the information being broadcast in wireless channel as data items. In the context of M-services, the broadcast data items can either be M-services or database records. In broadcast channel using (1, m) indexing, every broadcast data is indexed on its key attribute. For M-services, the key attribute is the service key of each M-service and for database it is the primary key of each database record. Indices are organized in B+ tree structure, which is referred to as *index tree*. The index and data organization of (1, m) indexing, which is shown in Figure 5.1, is very similar to that of distributed indexing (Figure 3.2).

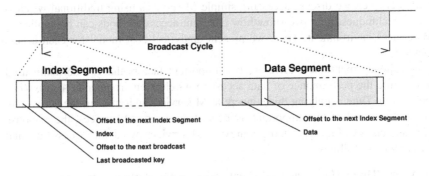

Fig. 5.1 Index and data organization of (1,m) indexing

In the original (1,m) indexing method, all broadcast buckets are of the same size. The bucket size is determined by the length of a broadcast data item (in this case, a M-service) plus an offset value. Each bucket may contain a few index entries depending on the size of data items and their primary keys. However, in the M-service channel, where M-services are much larger than the service keys, a bucket with the size equivalent to a M-service may contain many index entries up to the whole index tree. This obviously loses the flexibility of (1,m) indexing, because every time a mobile client needs to read index information, it has to download many unnecessary index entries or maybe even the whole tree. To preserve the flexibility of (1,m) indexing, for M-service channel, we use an improved (1,m) indexing method, in which index buckets have smaller size than M-services. The structure of the broadcast channel will still be the same as shown in Figure 5.1, except that the index buckets are smaller. Since the pointers stored in each index bucket are the absolute time offset to the child index nodes or data buckets, the size of the data buckets does not have any impact on the wireless channel structure.

5.1.2 Signature Indexing

We now discuss how the signature indexing method can be used to access M-service. In an M-service channel, the broadcast items are the M-services. Preceding each M-service, the corresponding service key is broadcast as the signature of that M-service. Mobile clients will read the service key first before downloading the whole M-service. Since the service key is unique, there will be no false drops if service key is used as signature directly. If the M-service size is fixed, the M-service channel structure will be the same as what is shown in Figure 3.3. When the sizes of M-services are different, an extra field indicating the size of the followed M-service bucket is needed in each signature bucket in the M-service channel. In other words, each signature bucket shown in Figure 3.3 will consist of two parts: the signature and the size of the M-service that follows. The size of the M-service here is expressed as the time offset to the next signature bucket. In the data channel, since we assume that all database records are equal in length, the data channel structure will be the same as in Figure 3.3.

5.1.3 Hashing

In M-service channel with fixed-size M-services and data channel, the broadcast channel will have the same structure as shown in Figure 3.4 because all broadcast items are equal in size. For M-service channel with varied-size M-services, the same method cannot be directly applied due to the fact that all M-services have different sizes. We propose an improved hashing method that can deal with varied-size broadcast items. In the improved hashing method, we use a small size broadcast bucket. As a result, every M-service may take one or more buckets to store. The broadcast channel is constructed as follows:

- Allocate N_m empty buckets, where N_m is the number of M-services.
- Generate a hash code for each M-service using the service key.
- Place each M-service in the allocated bucket based on its hash code.
- Calculate the number of buckets used to store this M-service and place the value in the first bucket of this M-service.
- For each M-service, consider the following scenarios:
  - If the M-service size is larger than the bucket size, keep creating new buckets until the M-service can fit in and shift all the rest of buckets forward.
  - When there is a hash code conflict, create one or more buckets to fit the M-service and shift the rest of buckets forward.
  - Change the shift value accordingly whenever a bucket is shifted forward.

There are two cases that can cause a bucket to shift forward: (1) when there is a hashing conflict, this is the same as in fixed size case; (2) when a M-service is larger than the bucket size. As a result, in the broadcast channel, all buckets will still be equal in size, except that some M-services may occupy several buckets. However, the structure of the broadcast channel is still the same as that in fixed size M-service channel. Therefore, the way of accessing M-service channel with varied-size M-services will be the same too.

Upon tuning into the broadcast channel, a mobile client will download the first complete bucket it comes across. Then the mobile client calculates the hash code of the request service key or primary key of the requested data item and compare the calculated hash code with the hash code stored in the current bucket to find the time offset to the right bucket (*hashing position*), which means the bucket with the matching hash code. The mobile client then goes to doze mode and wakes up when the right bucket arrives. If the shift value in that bucket is not empty, it means the actual bucket has been shifted because of collisions or varied-size M-services. The mobile client then goes to the bucket indicated by the shift value (*shift position*) to retrieve the requested service or data item.

5.2 Analytical Model

In this section, we present the analytical model for each access method. The total access time for each service request will be the sum of the time to obtain UDDI registry, download and execute the selected M-service, and retrieve the data items. The tuning time will be the time spent to actively listen to the M-service and data channels plus the time to execute the M-service. Table 5.1 defines the parameters and symbols used in this section. The access time and tuning time can be expressed as follows:

$$A_t = U_t + C_{tm} + A_{tm} + M_t + C_{td} + A_{td}$$
$$T_t = U_t + T_{tm} + M_t + T_{td}$$

To simplify the analysis, we assume that the time to obtain UDDI registry is a fixed value. It is also reasonable to assume that the switch time between two channels is a fixed value. Thus, in the above formulas, U_t, C_{tm}, M_t, and C_{td} are constants. Now what account for the access efficiency of each service request are the access time and tuning time of M-service and data channels. In the rest of this section, we will show how to derive access and tuning times for accessing M-service channel and data channel with different access methods.

System parameters	
N_m	Number of registered M-services
S_m	M-service program size
S_{mk}	Size of M-service key
N_o	Number of operations per service
N_r	Number of data items/records in database
S_{dk}	Key size of data items
S_d	Data item size
S_b	Logical broadcast unit (bucket) size
B_c	Broadcast cycle - the length of all contents in the broadcast channel
B_m	M-service wireless channel bandwidth
B_d	Database wireless channel bandwidth
Performance measurement parameters	
A_t	Total access time
T_t	Total tuning time
U_t	Time to obtain UDDI information
C_{tm}	Time to switch to M-service channel
A_{tm}	Access time to obtain M-service
T_{tm}	Tuning time to obtain M-service
M_t	Time to execute M-service
C_{td}	Time to switch to data channel
A_{td}	Access time to retrieve data items
T_{td}	Tuning time to retrieve data items
B_t	Broadcast cycle time - time to scan the whole broadcast channel

Table 5.1 Symbols and parameters for simple M-services

5.2.1 Accessing M-service channel

In this section, we investigate three different access methods that can be used to make M-service channel access more efficient, namely, *signature indexing*, *hashing*, and *B+ tree indexing*. We consider two scenarios for each method: (1) M-service channel contains fixed-size M-services; (2) M-service channel contains varied-size M-services.

5.2.1.1 No data access method

As a comparison benchmark, we present the performance of accessing M-service channel without access methods. Mobile clients need to traverse through the whole broadcast channel until the requested M-service is found. Therefore, the expected access and tuning times will not be dependent on whether the M-service size is fixed or not. When a mobile client tunes into a broadcast channel, it may hit any position of a broadcast bucket. The mobile client has to stay active until the first complete bucket is retrieved so as to acquire enough access information. We define the time spent for the first complete broadcast bucket to arrive as the *initial wait time* (F_t).

The average access time and tuning time are both half of the whole broadcast cycle time, plus the initial wait time:

$$A_{tm} = T_{tm} = \frac{1}{2} \times B_t + F_t$$
$$= \frac{1}{2} \times (N_m + 1) \times \frac{S_m}{B_m}$$

On average, the initial wait time is equal to the time to retrieve half of a bucket, which is $\frac{1}{2} \times \frac{S_m}{B_m}$. The initial wait time will be part of the access time and tuning time of every access method, but the calculation may vary.

5.2.1.2 M-service channel with fixed-size M-services

Signature indexing:

In this method, mobile clients always examine the preceding service key before downloading a M-service. Since the service key of each M-service is unique, there will be no false drop. The average access time is half of the broadcast cycle time plus the initial wait time. The tuning time consists of the initial wait time, time to retrieve $\frac{N_m}{2}$ signature buckets and download the requested M-service. The average initial wait time here is half of the time to scan a signature bucket and a M-service bucket, which is as $\frac{1}{2} \times (S_m + S_{mk})/B_m$. There are N_m signature buckets and N_m M-service buckets in a broadcast cycle. Thus, the broadcast cycle time is $N_m \times (S_m + S_{mk})/B_m$. Based on the above analysis, we have:

$$A_{tm} = \frac{1}{2} \times \frac{(S_m + S_{mk})}{B_m} + \frac{1}{2} \times N_m \times \frac{(S_m + S_{mk})}{B_m}$$
$$= \frac{1}{2} \times (N_m + 1) \times \frac{(S_m + S_{mk})}{B_m}$$
$$T_{tm} = \frac{1}{2} \times \frac{(S_m + S_{mk})}{B_m} + \frac{1}{2} \times N_m \times \frac{S_{mk}}{B_m} + \frac{S_m}{B_m}$$
$$= (1\frac{1}{2} \times S_m + \frac{1}{2} \times (N_m + 1) \times S_{mk})/B_m$$

Hashing

In this method, each service key is hashed to an integer value. The hashing function precedes each M-service to help locate the requested service key. The hash value of each service key also precedes the corresponding M-service. In cases of collisions, buckets in broadcast channel may be out of place. Therefore, an offset value is required at the beginning of each bucket to indicate the right position of the M-service with the correct hashing value. Let S_h be the hashing function size, S_{hk} be the hash value size, and S_{of} be the offset value size. Thus, the size of each bucket S_b is $S_m + S_h + S_{hk} + S_{of}$.

The access time of the hashing method consists of the initial wait time (F_t), time to reach the *hashing position* (H_t), time to reach the *shift position* (S_t), time to retrieve colliding buckets (C_t), and time to download the required bucket (D_t). D_t is the time to read one complete bucket, which is $\frac{S_b}{B_m}$. Let N_c be the number of colliding buckets, if the collisions are uniformly distributed among all M-services, the average number of shifts for each bucket is thus $\frac{N_c}{2}$. Therefore, we have $S_t = \frac{N_c}{2} \times \frac{S_b}{B_m}$. Furthermore, the average number of colliding buckets for each hashing value is $\frac{N_c}{N_m}$. Thus, we have $C_t = \frac{N_c}{N_m} \times \frac{S_b}{B_m}$. There is more involved in the calculation of H_t. Assume the number of initially allocated buckets is the number of M-services N_m and the number of colliding buckets is N_c. The resulting total number of buckets in the broadcast cycle is $N = N_m + N_c$. We calculate H_t based on the position of the first arriving bucket and whether the requested information has been broadcast or not. Assuming that the position of the first arriving bucket is n, H_t consists of the following three parts:

$$H_{ti} = \begin{cases} H_{t1} & (n > N_m) \\ H_{t2} & (n \le N_m \text{ and } req\_item\_broadcast = False) \\ H_{t3} & (n \le N_m \text{ and } req\_item\_broadcast = True) \end{cases}$$

The *req_item_broadcast* above designates if the requested information has already been broadcast in the current broadcast cycle. Each part of H_t is derived as follows:

$$H_{t1} = \frac{N_c}{N} \times (\frac{1}{2} \times (N_c + N_m)) \times \frac{S_b}{B_m} = \frac{1}{2} \times N_c \times \frac{S_b}{B_m}$$

$$H_{t2} = \frac{1}{2} \times \frac{N_m}{N} \times \frac{N_m}{3} \times \frac{S_b}{B_m}$$

$$H_{t3} = \frac{1}{2} \times \frac{N_m}{N} \times (\frac{N_m}{3} + N_c + \frac{N_m}{3}) \times \frac{S_b}{B_m}$$

The first part of each formula above is the probability the scenario will happen. As a result, we have $H_t = H_{t1} + H_{t2} + H_{t3}$. Based on the above discussion, the access time can be derived as:

$$A_{tm} = F_t + H_t + S_t + C_t + D_t$$
$$= (N_c + \frac{1}{2} \times N_m + \frac{N_c}{N_m} + 1\frac{1}{2}) \times \frac{(S_m + S_h + S_{hk} + S_{of})}{B_m}$$

The tuning time consists of the initial wait time, the time to read the first bucket to obtain the hashing position (read one bucket), time to obtain the shift position (read one bucket), and time to retrieve the colliding buckets (C_t), and time to download the required bucket (read one bucket). The probability of collision is $\frac{N_c}{N_m}$. Thus, we have $C_t = \frac{N_c}{N_m} \times S_b/B_m$. For those requests that tune in at the time which the requested bucket has already been broadcast, one extra bucket read is needed to start from the beginning of the next broadcast cycle. The probability of this scenario occurrence is $(N_c + \frac{1}{2} \times N_m)/(N_c + N_m)$. As a result, the expected tuning time is:

$$T_{tm} = (\frac{N_c + \frac{1}{2} \times N_m}{N_c + N_m} + \frac{N_c}{N_m} + 3\frac{1}{2}) \times \frac{S_b}{B_m}$$

$$= (\frac{N_c + \frac{1}{2} \times N_m}{N_c + N_m} + \frac{N_c}{N_m} + 3\frac{1}{2}) \times \frac{(S_m + S_h + S_{hk} + S_{of})}{B_m}$$

(1,m) indexing

Let n be the number of index entries can be stored in a bucket, intuitively $n = \lfloor \frac{S_m}{S_{mk}} \rfloor$. Let k be the number of levels of the index tree. It is also intuitive that $k = \lceil \log_n(N_m) \rceil$. The access time consists of three parts: the initial wait time, *initial index probe time*, and *broadcast wait time*. Let S_{ib} be the size of index bucket and S_{mb} be the size of M-service buckets. We have the following derivations:

initial wait (F_t): The calculation of the initial wait is different from that of signature indexing because the number of index buckets is not the same as the number of M-service buckets. When a client tunes into the broadcast channel, it may hit an index or a data bucket arbitrarily. Let N_i be the number of index buckets in a broadcast cycle, the probability of hitting a index bucket is $\frac{m \times N_i \times S_{ib}}{B_c}$ and M-service bucket is $\frac{N_m \times S_{mb}}{B_c}$. Therefore, the initial wait is:

$$F_t = \frac{1}{2} \times (\frac{m \times N_i \times S_{ib}}{B_c} \times S_{ib} + \frac{N_m \times S_{mb}}{B_c} \times S_{mb})/B_m$$

initial index probe (P_t): This part is the time to reach the first index segment. It can be expressed as the average time to reach the next index segment, which is calculated as the sum of the average length of index segments and data segments. With an n-ary index tree of k levels, the number of index buckets (N_i) is:

$$N_i = 1 + n + \dots + n^{k-1} = \frac{n^k - 1}{n - 1}$$

where $k = \lceil \log_{\lfloor \frac{S_{ib}}{S_{mk}} \rfloor}(N_m) \rceil$. The average number of data buckets in a data segment is $\frac{N_m}{m}$. Therefore, the initial index probe is calculated as:

$$P_t = \frac{1}{2} \times \left(\frac{n^k - 1}{n - 1} \times S_{ib} + \frac{N_m}{m} \times S_{mb}\right)/B_m$$

broadcast wait (W_t): This is the time from reaching the first indexing segment to finding the requested M-service. It is approximately half of the whole broadcast cycle time B_c, which is $(m \times N_i \times S_{ib} + N_m \times S_{mb})/B_m$. Thus, the total access time is:

$$A_{tm} = F_t + P_t + W_t$$

The tuning time is easier to calculate than access time, because during most of the probes clients are in doze mode. It includes the initial wait F_t, the time to read the first bucket to find the first index segment (read one index or data bucket, which is $2 \times F_t$), the time to traverse the index tree (read k buckets), and the time to download the M-service (read one bucket). Thus, the tuning time can be derived as:

$$T_{tm} = F_t + 2 \times F_t + (k \times S_{ib} + S_{mb})/B_m$$
$$= 3 \times F_t + \frac{k \times S_{ib} + S_{mb}}{B_m}$$

5.2.1.3 M-service channel with varied-size M-services

Signature indexing:

Since the M-services have different sizes, in front of each M-service we also need the size of the M-service to direct the mobile clients to the next M-service. Assume the size of the variable containing M-service size is S_{ms}, we have

$$A_{tm} = \frac{1}{2} \times (N_m + 1) \times \frac{S_m + S_{mk} + S_{ms}}{B_m}$$
$$T_{tm} = \frac{\frac{1}{2} \times (S_m + S_{mk} + S_{ms}) + \frac{1}{2} \times N_m \times (S_{mk} + S_{ms}) + S_m}{B_m}$$
$$= \frac{1\frac{1}{2} \times S_m + \frac{1}{2} \times (N_m + 1) \times (S_{mk} + S_{ms})}{B_m}$$

Hashing

The analysis process of the analytical model for hashing in case of varied M-service size will be the same as that of fixed M-service size. All M-services will still be hashed into different positions in the broadcast channel by using a hash function.

However, each M-service may need one or more buckets to store it. Furthermore, now we have two reasons that could cause M-services to be out of place: the large size of M-services and the collisions. Let N_{co} be the sum of conflict and offset items, $\bar{n}$ be the average number of buckets per service. Following the same analysis process in Section 5.2.1, we can derive the access time and tuning time as follows:

$$F_t = \frac{1}{2} \times \frac{S_b}{B_m}$$

$$H_{t1} = \frac{N_{co}}{N} \times (\frac{1}{2} \times (N_{co} + N_m)) \times \frac{S_b}{B_m}$$

$$H_{t2} = \frac{1}{2} \times \frac{N_m}{N} \times \frac{N_m}{3} \times \frac{S_b}{B_m}$$

$$H_{t3} = \frac{1}{2} \times \frac{N_m}{N} \times (\frac{N_m}{3} + N_c + \frac{N_m}{3}) \times \frac{S_b}{B_m}$$

$$S_t = \frac{N_{co}}{2} \times \frac{S_b}{B_m}$$

$$C_t = \frac{N_c}{N_m} \times \frac{S_b}{B_m}$$

$$D_t = \bar{n} \times \frac{S_b}{B_m}$$

$$A_{tm} = F_t + H_t + S_t + C_t + D_t$$

$$= (N_{co} + \frac{1}{2}\frac{N_m^2}{N} + \frac{1}{2}\frac{N_m \times N_c}{N} + \frac{N_c}{N}_m + \bar{n} + \frac{1}{2}) \times S_b/B_m$$

$$T_{tm} = (\frac{N_{co} + \frac{1}{2} \times N_m}{N_c + N_m} + (\frac{N_c}{N_m} + 1) \times \bar{n} + 2\frac{1}{2}) \times \frac{S_b}{B_m}$$

(1,m) indexing

The analytical model of B+ tree indexing for varied-size M-service channel is the same as the one for fixed-size M-service channel. The time offset values contained in each bucket are absolute values. The differences in the M-services sizes only result in the differences in the values of those time offsets. The expressions of the access time and tuning time are still the same as those presented in Section 5.2.1.

5.2.2 Accessing data channel

In this section, we present the analytical model for each access method being applied to the data channel when each request acquires one or more data items. Let N_o be the number of operations per mobile service and N_{do} be the number of data items requested by an operation. The total number of data items requested per M-service is thus $N_o \times N_{do}$. Without losing the generality, to simplify the analysis, we assume that every operation only requests for one data item. Therefore, for each M-service, N_o data items will be requested.

5.2.2.1 No data access method

In this case, no access method is used to access data in the data channel. Mobile clients need to traverse through the whole broadcast channel until all the requested N_o data items are found. If all data items are uniformly distributed in the data channel, the average traversing distance to reach the last requested data item will be $\frac{N_o}{N_o+1}$ of the whole broadcast cycle time plus the initial wait time. Thus, we have

$$A_{td} = T_{td} = \frac{N_o}{N_o+1} \times B_t + F_t$$

$$= (\frac{N_o}{N_o+1} \times N_r + \frac{1}{2}) \times \frac{S_d}{B_d}$$

5.2.2.2 Signature indexing

Mobile clients always examine the primary key first before downloading the whole data item. To find all requested data items, mobile clients need to finish listening to $\frac{N_o}{N_o+1}$ of the whole broadcast cycle. The access time and tuning time can be derived as follows:

$$A_{td} = (\frac{N_o}{N_o+1} \times N_r + \frac{1}{2}) \times \frac{S_d + S_{dk}}{B_d}$$

$$T_{td} = \frac{\frac{1}{2} \times (S_d + S_{dk}) + \frac{N_o}{N_o+1} \times N_r \times S_{dk} + N_o \times S_d}{B_d}$$

5.2.2.3 Hashing

The access time and tuning time of hashing for data channel can be derived based on the analysis in Section 5.2.1. The only difference is that for data channel, N_o data items are acquired for each request. When tuning into the data channel, a mobile client downloads the first complete bucket to obtain the hashing function and the hash code of the current bucket. Then the mobile client calculates the hash code for each request key and compares it with the hash code stored in the current bucket. From the comparisons, the distance to the hashing position of each data item can be derived. The mobile client then saves these distance values in a distance list in the order of the distances and goes to them one by one following the distance list. After reaching each hashing position, the mobile client follows the same procedure as described in Section 5.2.1 to find the requested data item. At each step, the distance list is updated with any new distance value obtained. Since all these steps are done sequentially, the total access time to retrieve all N_o data items will be the access time spent to retrieve the farthest data item from the point the mobile client tunes in. The same applies to the tuning time. With this in mind, our analysis can be simplified as requesting only one data item given that the item is the farthest one of N_o uniformly distributed data items from the point a mobile client tunes in.

Based in the analysis in Section 5.2.1, the access time still consists of: F_t, H_t, S_t, and C_t, plus the time to download the requested data items (read N_o buckets). If the position of the first arriving bucket is after N_r, the average time to get the farthest data item will be $\frac{N_o}{N_o+1} \times (N_c + N_r) \times \frac{S_b}{B_d}$; if the first arriving bucket is in N_r, and any of the requested items is not broadcast yet, the average time to get the farthest data item will be $\frac{N_o}{N_o+2} \times N_r \times \frac{S_b}{B_d}$; if the first arriving bucket is in N_r, and one of the requested items is not broadcast yet, the average time to get the farthest data item will be $(\frac{N_o+1}{N_o+2} \times N_r + N_c) \times \frac{S_b}{B_d}$. With the special consideration of the probability, each of this time can be derived as follows:

$$H_{t1} = \frac{N_c}{N} \times (\frac{N_o}{N_o+1} \times (N_c+N_r)) = \frac{N_o}{N_o+1} \times N_c \times \frac{S_b}{B_d}$$

$$H_{t2} = \frac{1}{N_o+1} \times \frac{N_r}{N} \times \frac{N_o}{N_o+2} \times N_r \times \frac{S_b}{B_d}$$

$$H_{t3} = \frac{N_o}{N_o+1} \times \frac{N_r}{N} \times (\frac{N_o+1}{N_o+2} \times N_r + N_c) \times \frac{S_b}{B_d}$$

$$S_t = \frac{1}{2} \times N_c \times \frac{S_b}{B_d}$$

$$C_t = \frac{N_o}{2} \times \frac{N_c}{N_r} \times \frac{S_b}{B_d}$$

$$D_t = N_o \times \frac{S_b}{B_d}$$

$$F_t = \frac{1}{2} \times \frac{S_b}{B_d}$$

The bucket size S_b above is equal to $S_d + S_h + S_{hk} + S_{of}$. The access time can be derived from the above formulas as follows:

$$A_{tm} = H_t + S_t + C_t + D_t + F_t$$

The tuning time consists of the initial wait time, the time to read the first bucket to obtain the hashing positions for all requested data items (read one bucket), time to obtain the shift position (read N_o bucket), and time to retrieve the colliding buckets

$(C_t \times N_o)$, and time to download the required bucket (read N_o buckets). The probability of collision is $\frac{N_c}{N_r}$. Thus, we have $C_t = \frac{N_c}{N_r} \times \frac{S_b}{B_d}$. For those requests that tune in at the time when the farthest requested bucket has already been broadcast, one extra bucket read is needed to start from the beginning of the next broadcast cycle. The probability of this scenario occurrence is $((N_c + \frac{N_o}{N_o+1} \times N_r)/(N_c + N_r)) \times \frac{S_b}{B_d}$. As a result, considering the initial wait time, the expected tuning time is:

$$T_{tm} = (1\frac{1}{2} + N_o + \frac{N_c}{N_r} \times N_o + N_o + \frac{N_c + \frac{N_o}{N_o+1} \times N_r}{N_c + N_r}) \times \frac{S_b}{B_d}$$

5.2.2.4 (1,m) indexing

When requesting for multiple data items (N_o), we need a local list to save a sequence of time offset values to the buckets to be visited next. Using the index tree shown in Figure 3.1 as an example, we illustrate how multiple data items are retrieved by a request. Assume we are requesting for data items 12 and 66. The following steps are taken to retrieve these two data items:

- Mobile client tunes into data channel.
- Goes to doze mode and wakes up at the beginning of the closest index segment.
- Compares the keys of the requested data items (12 and 66) with the index entries in the root node/bucket. And saves the time offset values to a1 and a3 in a local list with a1 in front because a1 is closer.
- Goes to bucket containing node a1 and compares the keys again. Then saves b2 in the list with the new sequence a3, b2 because a3 is closer. Offset value to a1 is removed because it is already visited.
- Goes to bucket containing a3 and compares the keys. Removes a3 and saves b8 in the list with the sequence b2, b8.
- Goes to b2, compares the keys, and saves c5 after b8 in the list.
- Goes to b8, compares the keys, and saves c23 after c5 in the list.
- Goes to bucket containing data item 12 and download the data item.
- Goes to bucket containing data item 66 and download the data item.

Following the above procedure, the access time is equal to the time spent to retrieve the data item pointed by the index entry stored in the right most index bucket in the index tree. Based on the analysis in Section 5.2.1, the access time consists of the initial wait time (F_t), the initial index probe (P_t) and the broadcast wait (W_t). Let S_{db} be the size of data buckets. Since the index and data buckets have same size in data channel, the initial wait time can be simplified as follows:

$$F_t = \frac{1}{2} \times \frac{S_{db}}{B_d}$$

The derivation of initial index probe is unchanged, which is:

$$P_t = \frac{1}{2} \times \left(\frac{n^k - 1}{n - 1} + \frac{N_r}{m}\right) \times \frac{S_{db}}{B_d}$$

The broadcast wait will be approximately $\frac{N_o}{N_o+1}$ of the whole broadcast cycle, which is $\frac{N_o}{N_o+1} \times N_r \times \frac{S_{db}}{B_d}$. Thus, the total access time is:

$$
\begin{aligned}
A_{td} &= F_t + P_t + W_t \\
&= \left(\frac{1}{2} + \frac{n^k - 1}{2(n-1)} + \frac{N_r}{2m} + \frac{N_o \times N_r}{N_o + 1}\right) \times \frac{S_{db}}{B_d}
\end{aligned}
$$

where $k = \left\lceil \log_{\lfloor \frac{S_d}{S_{dk}} \rfloor}(N_r) \right\rceil$.

The tuning time includes the initial wait time F_t, the time to read the first bucket to find the first index segment (read one bucket), the time T_i to traverse the index tree for N_o data items, and the time T_d to download all N_o data items, which is $N_o \times \frac{S_{db}}{B_d}$. It is hard to derive the exact value of T_i because it depends on the distribution of the N_o requested data items in the index tree. Instead, we derive the average value of T_i. The minimum value of T_i is achieved when all N_o data items are siblings in the index tree, in which K index bucket probes are required. The value of T_i reaches maximum when all N_o requested data items are distributed uniformly in the index tree. In this case we need $\log_{\lceil \frac{n}{2} \rceil}(N_m)$ probes for each data item. Therefore, the average T_i is:

$$T_i(avg) = \frac{K + N_o \times \log_{\lceil \frac{n}{2} \rceil}(N_o)}{2} \times \frac{S_{db}}{B_d}$$

As a result, the tuning time is:

$$T_{td} = F_t + P_t + T_i(avg) + T_d$$

5.3 Practical Study

In this section, we present our experiments to evaluate the performance of the proposed methods. The experiments are performed using an adaptive testbed [69] developed to simulate wireless data access. We further enhanced the testbed to simulate an M-service environment and to be capable of supporting fixed-size and varied-size M-service channel. For the same reasons as mentioned in Section 3.4, all the analytical and simulation results are presented in the form of the length of the broadcast data that has been traversed, instead of the actual time spent. The access time for each method is represented by the length of all broadcast buckets passed by when requesting an M-service or data item. The tuning time of a request is calculated as the length of all buckets actively accessed (listened and downloaded) by the request.

We assume that every service request starts with requesting for an M-service, and after the execution of the mobile service, the mobile client will start listening to the data channel until the requested data item(s) are found and downloaded. The mobile client does not have to go back and forth between the mobile services channel and data channel. Based on this assumption, we consider the performance measurement of accessing mobile services and accessing data items to be two sequential processes and can be analyzed separately.

The experiments consist of three cases: (1) accessing fixed-size M-service channel; (2) accessing varied-size M-service channel; (3) accessing data channel. For each case, we present the simulation settings we used to evaluate the M-services and data access. The outcome of each experiment will be measured by access and tuning times.

Mobile service key size (S_{mk})	16 bytes
Data record key size (S_{dk})	10 bytes
Data record size (S_d)	100 bytes
Number of requests	> 10,000
Confidence level	0.95
Confidence accuracy	0.05
Requests interval	exponential distribution
Requests generation distribution	uniform distribution

Table 5.2 Simulation settings for simple M-services

Table 5.2 lists all common simulation settings used in our experiments. In this proposal, we assume that all mobile services or data items have equal probability of being accessed, which means the requested mobile services and data items in our simulation are selected following uniform distribution. We use the standard mobile service key size, which is 16 bytes [61].

5.3.1 Performance measurement of accessing fixed-size mobile services channel

Table 5.3 lists all the experiments we conducted for accessing fixed-size mobile services channel and their simulation settings.

Figure 5.2 shows the simulation results for accessing fixed-size mobile services channel using plain broadcast, (1,m) indexing, signature indexing and hashing. The lines marked with (S) are simulation results. Those marked with (A) are analytical results. We observe in both figures that the simulation results match the analytical results very well.

	Access Method	N_m	S_m
Experiment 1	plain	*1 - 1000*	50 KB
Experiment 2	Signature indexing	*1 - 1000*	50 KB
Experiment 3	Hashing	*1 - 1000*	50 KB
Experiment 4	B+ tree indexing	*1 - 1000*	50 KB

Table 5.3 Accessing broadcast channel with fixed-size mobile services

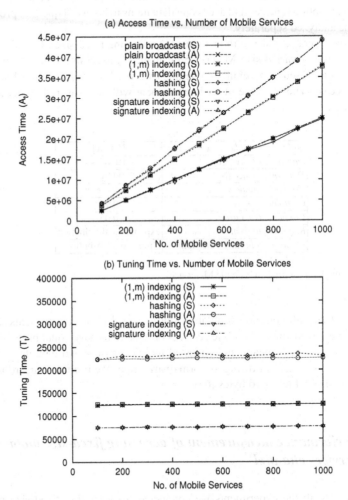

Fig. 5.2 Compare all access methods for fixed size M-services channel

As shown in Figure 5.2, signature indexing exhibits the best performance in both access time and tuning time. The reason is that the size of the mobile services is much bigger than that of the service key, which is used as the signature for each mobile service. Therefore, the overhead introduced by adding signature for each mobile service in the broadcast channel is very small compared to the whole broadcast cycle. This results in an access time close to the plain broadcast (as shown in Figure 5.2), which has the optimum access time performance. The tuning time of signature indexing is determined by both signature size and the number of false drops. As we mentioned in Section 5.2.1, since all service keys are unique, there is no false drop for signature indexing. So the tuning time of signature indexing consists of the reading time of a series of signatures (looking for a match) and the time to download the requested mobile service. This value is smaller than other methods because the signature size is considerably smaller than M-service size.

To help us better understand the performance trends shown in the figures, we define two terms here, *access method overhead* and *conflict overhead*. The access method overhead means the overhead introduced to the broadcast cycle to apply access methods to the broadcast channel, such as hashing values, hashing functions, signatures, and indices. And the conflict overhead designates the overhead produced by data conflicts when applying access methods to broadcast channel. Examples of conflict overhead are false drops in signature indexing and hashing conflicts in hashing method. It is intuitive that the access method overhead is determined by the number and size of the extra information required by different access methods. For example, the access method overhead in (1,m) indexing is determined by the number of index buckets in the index tree and the number of segments. On the other hand, the conflict overhead is usually determined by the conflict rate and the size of the broadcast data, in this case, the mobile services. In the experiments covered by this section, the conflict overhead is much greater than the access method overhead because the size of M-services is a lot larger than that of service keys, which are used as signatures in signature indexing and primary key in (1,m) indexing. Since hashing is the only method that may have conflict overhead, it has the worst performance in both access and tuning times, which is also proved by the simulation results.

The performance of (1,m) indexing falls in the middle. The reasons are, on one hand, (1,m) indexing does not have conflict overhead, it thus has better performance than hashing. On the other hand, it needs to broadcast index tree m times in a broadcast cycle, which results in a greater access method overhead. Therefore, the resulting performance is not as good as signature indexing.

As a result, it is obvious that signature indexing is the most suitable access method for accessing broadcast channel with fixed-size mobile services.

5.3.2 Performance measurement of accessing varied-size mobile services channel

In this section, we present the simulation experiments we conducted for accessing broadcast channel that contains varied-size M-services. In these experiments, we define a range for size of M-services being broadcast and assume that the sizes follow uniform distribution in that range. Table 5.4 shows the simulation settings. The simulation and analytical results for different methods are presented in Figure 5.3. Again, we observe that the simulation results match the analytical results very well.

	Access Method	N_m	S_m
Experiment 6	None	*1 - 1000*	1 - 100 KB
Experiment 7	Signature indexing	*1 - 1000*	1 - 100 KB
Experiment 8	Hashing	*1 - 1000*	1 - 100 KB
Experiment 9	B+ tree indexing	*1 - 1000*	1 - 100 KB

Table 5.4 Accessing broadcast channel with varied-size mobile services

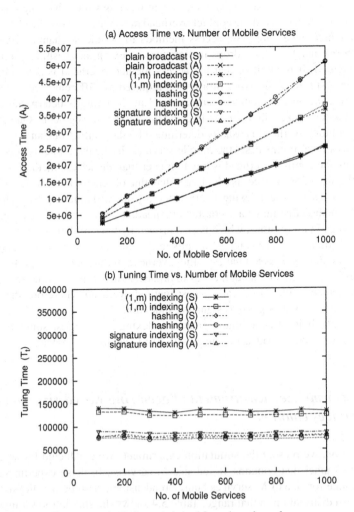

Fig. 5.3 Compare all access methods for varied size M-services channel

The access time performance shown in Figure 5.3(a) is very similar to that in accessing fixed-size M-service channel. The only difference is that the values are slightly larger. This is because in each method (except plain broadcast), extra information is broadcast to store the size of every M-service/bucket. (1,m) indexing and signature indexing exhibit similar performance trends in tuning time as well. However, we observe a great tuning time improvement for hashing method. This improvement is caused by the technique we used to apply hashing to varied-size M-service channel. Since hashing methods requires all buckets to the the same size, we cannot directly put every M-service in a single bucket and place the in the broadcast channel. As introduced in Section 5.2.1, we use a small size bucket as the basic broadcast unit. Every M-service may take up one or more buckets to store. By using this technique, we can still take advantage of hashing method in varied-size M-service channel. Another improvement achieved by this technique on hashing method is the improved tuning time. The conflict overhead on tuning time is now determined by the bucket size instead of the M-service size because a mobile client only needs to read the first bucket of an M-service to find out if it is the requested M-service. If not, with the help of the extra information stored in the broadcast channel indicating the size of the current M-service, the mobile client will go to doze mode and wake up when the next M-service arrives. Since the bucket size now is much smaller than the average M-service sizes, the resulting tuning time will be consequently much smaller too. Even though the hashing method exhibits similar and sometimes even better tuning time performance than the signature indexing, the latter is still preferred in varied-size M-service channel because of its good performance in access time. Therefore, we conclude that the signature indexing is the most suitable access method for M-services channel, either with fixed-size or varied-size M-services.

5.3.3 Performance measurement of accessing data channel

As a result of executing M-services, mobile clients will tune into data channel to request data items. We conducted extensive experimental study on performance evaluation of accessing data channel for different access methods [69]. However, all the completed experiments assume that a mobile client only requests for one data item at a time. This is obvious not applicable in the M-service environment, where each M-service may have one or more operations, and each of such operation may request one or more data items. As discussed in Section 5.2.2, without losing generality, we assume that each operation requests one data item only. Therefore, each M-service requests N_o data items, where N_o is the number of operations contained in the M-service.

In this section, we present the simulation experiments for accessing data channel with multiple operations for each M-service. We consider two scenarios: (1) every M-service has fixed number of operations; and (2) every M-service has varied number of operations.

5.3.3.1 Accessing data channel with fixed number of operations

Table 5.5 lists all simulation experiments we conducted for the scenario that each M-service has 5 operations. We assume that the data channel may contain up to 100,000 data items.

	Access Method	N_o	N_d
Experiment 11	None	5	*10,000 - 100,000*
Experiment 12	Hashing	5	*10,000 - 100,000*
Experiment 13	B+ tree indexing	5	*10,000 - 100,000*
Experiment 14	Signature indexing	5	*10,000 - 100,000*

Table 5.5 Accessing data channel with fixed number of operations

We observe from Figure 5.4 that plain broadcast and signature still have the best access time performance which is consistent with our previous study [69]. However, we also observe that hashing method exhibits better performance than (1,m) indexing, which is contradictory to the results of the simulation experiments we conducted before, where only one data item is acquired by a request. This phenomenon can be explained as follows: With (1,m) indexing, the whole index tree is broadcast m times in each broadcast cycle. When only one data item is requested, on average a request will traverse though about half of these trees. But when N_o data items are requested, a request will have to traverse through most of these trees depending on the value of N_o. This overhead results in a larger access time for (1,m) indexing.

Figure 5.4 only shows the tuning time for (1,m) indexing and hashing method because the tuning time for plain broadcast and signature indexing are too large to fit into this figure. We therefore exclude them from our analysis. As we can see from Figure 5.4, hashing method still has the best tuning time among all methods, which is also consistent with our previous study.

5.3.3.2 Accessing data channel with varied number of operations

We now consider the scenario that each M-service has different number of operations. This scenario is closer to the real world. We assume that the data channel contains 50,000 data items. Table 5.6 lists all the experiment settings.

	Access Method	N_o	N_d
Experiment 15	None	*1 - 10*	50,000
Experiment 16	Hashing	*1 - 10*	50,000
Experiment 17	B+ tree indexing	*1 - 10*	50,000
Experiment 18	Signature indexing	*1 - 10*	50,000

Table 5.6 Accessing data channel with varied number of operations

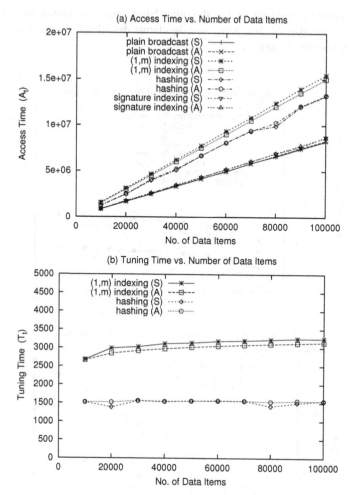

Fig. 5.4 Compare all access methods for data channel with fixed number of operations

Figure 5.5 shows the simulation results for accessing the data channel with varied number of operations. We observe that both access time and tuning time of every access method increase with the number of operations per M-service. It is noticeable that the increase of access time is not linear. The increase becomes slower when the number of operations gets larger. This can be explained by using plain broadcast as an example. When requesting one data item in plain broadcast, a mobile client needs to traverse averagely half of the broadcast cycle to obtain the data item. With two data items, the mobile client needs to go through two thirds of the broadcast cycle on average, and three fourth of the broadcast cycle for three data items, four fifth for four, and so on. As the number of data items increases, the *increase* value of the traverse length decreases, which indicates the access time will increase more slowly with the number of operations. On the other hand, since the tuning time directly relates to the number of data items mobile clients download, it increases linearly with the number of operations.

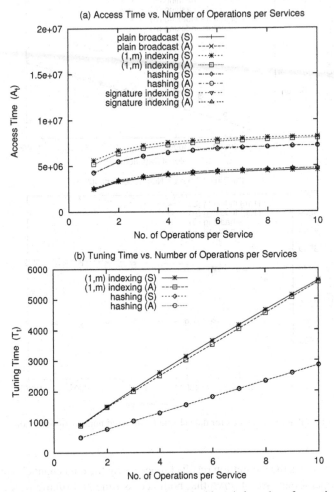

Fig. 5.5 Compare all access methods for data channel with varied number of operations

Comparing the access methods, we still see the same performance trends as in the last section. Plain broadcast and signature indexing have the best access time performance. Hashing method still has the best tuning time performance and has better access time performance than (1,m) indexing.

5.3.4 *Summary*

From the experiments, we note the following observations: (1) Plain broadcast (no access method) has the best access time but the worst tuning time. Since the tuning time of plain broadcast is far larger than that of any access method, it is usually not a preferred method in power limited wireless environments; (2) Signature indexing is the most suitable access method for M-service channel since it exhibits better access time and tuning time performance than all other methods; (3) When broadcast information has very large size (e.g. M-services), using smaller broadcast buckets can increase the performance of hashing method; (4) Access methods can be applied to varied-size M-service channel without introducing considerable overhead. (5) In data channel, signature indexing achieves better access time than most other access methods. However, the tuning time of signature indexing is pretty large. When energy is of little concern compared to waiting time, signature indexing is a preferred method. (6) Hashing achieves the best tuning time and good access time in data channel. For energy critical applications or at least when energy is not a negligible factor, hashing method will be the preferred method in data channel. The experiments showed that the best techniques for M-services and data channels are signature indexing and hashing methods respectively.

Chapter 6
Semantic Access to Composite M-services

In this chapter, we study how to efficiently access composite M-services in wireless broadcast systems. First, we present a novel wireless broadcast infrastructure that supports discovery and composition of M-services. Then we define access semantics for this infrastructure and study how to leverage these semantics to achieve best possible access efficiency.

6.1 Broadcast-based M-services Infrastructure

In this section, we propose a broadcast based composite M-services infrastructure. The main challenges for supporting composite M-services in wireless broadcast systems are as follows:

Service discovery: For users to be able to use the provided services, they must know what services are available and how to find suitable services to satisfy their needs. Typical service discovery strategies are interaction oriented. Users usually need to browse through the service registry, which is located on the server side, to find suitable services. In a broadcast-based environment, users passively listen to broadcast channels and filter on the received information. There would not be any active interaction between mobile users and wireless servers. New service discovery strategies that are suitable for broadcast-based environments need to be developed.

Efficient access to M-services and wireless data: The purpose of the proposed infrastructure is for mobile users to discover, access, and execute M-services and then retrieve wireless data when needed. So the main challenges for the infrastructure are 1) how to organize M-services and wireless data in broadcast channels for mobile users to access; and 2) how to efficiently access the services and data based on the channel organization.

Service composition: It is quite often a user needs to ask for multiple services in order to fulfill a request. In this case, *service composition* is required. Service composition refers to the aggregation of multiple services to create business processes that meet different needs. Typically, composite services are pre-defined and

X. Yang and A. Bouguettaya, *Access to Mobile Services*, Advances in Database Systems 38,
DOI: 10.1007/978-0-387-88755-5_6, © Springer Science + Business Media, LLC 2009

directly accessible by end users. The key challenges with service composition in broadcast-based environment are how to define composite services and how to efficiently deliver them to mobile users.

In the rest of this section, we present a broadcast-based M-services infrastructure that addresses the above listed challenges.

6.1.1 System Roles

An M-services system consists of three basic roles, *service providers*, *wireless carriers* and *mobile users*. Figure 6.1 illustrates the relationships between these roles.

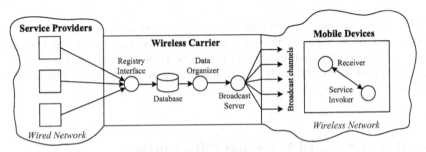

Fig. 6.1 System Roles

Service Providers

Service providers are the parties that provide M-services and wireless data to end users. They provide these services through wireless carriers who own wireless networks. The interaction between the service providers and wireless carriers takes place through wired networks such as the Internet or LANs. The following operations are performed by service providers:

- *Publish services*: Service providers first need to publish their services. Using established interfaces over wired networks, service providers send registry information of each service and the actual service code to the wireless carrier. They may also send service composition information for any provided composite services to the wireless carrier.
- *Provide live data feeds*: Service providers are also responsible for providing live data feeds to the wireless carrier. These data feeds are used by the provided services.
- *Provide service updates*: When a service is updated, its service provider needs to send updated registry information and the actual service to the wireless carrier.

A service provider may also send a deregister request to the wireless carrier to have a service removed from the broadcast channels.

Wireless Carrier

Wireless carriers are business entities that provide wireless communication services to users. They own wireless networks. They usually sell mobile devices and service plans to end users for them to access wireless networks. Typical examples of wireless carriers include *Verizon wireless, Cingular Wireless, T-Mobile,* and etc. They are responsible for organizing services and data on broadcast channels and make them available to mobile users. A wireless carrier has the following responsibilities:

- *Defining and providing service provider interfaces*: A wireless carrier needs to define interfaces for service providers to publish and update services, and send live data feeds.
- *Providing service registry*: A local registry needs to be provided by a wireless carrier to save published services and their registry information.
- *Organizing and broadcasting services and data*: A wireless carrier also needs to make all published information available on broadcast channels for mobile users to access. The information includes service registry information, actual service executables, composite services, and wireless data sets. The carrier first needs to define a channel organization that is easy for users to find and retrieve all the broadcast information. Then it needs to periodically broadcast the information based on the defined organization.

Mobile Users

Mobile users are the end users for those services provided by the service providers. Mobile users are equipped with mobile devices, such as cell phones or palms, which can access broadcast channels to retrieve M-services and wireless data. In the proposed infrastructure, a mobile device needs to have the following capabilities:

- *Wireless receivers*: First of all, a mobile device needs to be equipped with one or more wireless receivers that can listen to the broadcast channels and filter information based on user input.
- *Service invokers*: The infrastructure defines a generic service invoker that is used to execute retrieved services. The service invoker supports the following functions:

  - Interpret service registry or composite service records and prompt for user input based on the definition in the "¡input¿" element.
  - Tell wireless receivers what services need to be retrieved based on the registry or composite service records.
  - Invoke received services.

- When needed, pass the output of a service as the input to the next service for a sequential invocation sequence.

6.1.2 Interaction Model

Let us now take a closer look at how different system roles interact with each other. Figure 6.2 shows the interaction model for broadcast based M-services. The model aims to support basic Web services operations in wireless broadcast systems, such as service discovery, access to services, service execution and data retrieval.

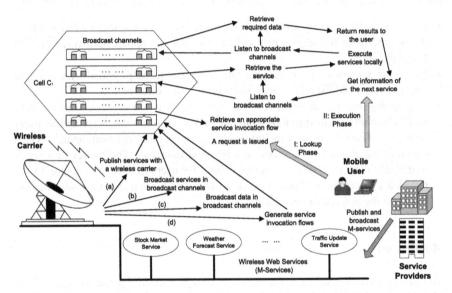

Fig. 6.2 Interaction Model

As can be seen from Figure 6.2, a service provider first publishes its services with a wireless carrier. It then broadcasts the actual services and required data via the wireless carrier's network. On the other hand, a mobile user first subscribes to services of his/her interest with the service provider. Then the user can access these services using his/her mobile devices from broadcast channels via the wireless network. The requested services are retrieved to the mobile device and executed locally. At last, upon the execution of the services, wireless data is retrieved from broadcast channels to fulfill the user request. We present as follows an algorithm with the main steps that need to be performed to execute a composite M-service in the proposed infrastructure.

```
Algorithm execute-M-service
```

```
/* executed whenever a mobile client wants to */
/* access an M-service in the local area */
Begin
    1: lookup M-services through registry channel.
    2: select an M-service.
    3: retrieve the description of the service via
       description channel.
    4: select the M-service operation that needs to
       be invoked.
    5: determine the input parameters required by
       that operation from the description.
    6: listen to broadcast channels in the local
       geographic area.
    7: download the executable of the selected
       M-service.
    8: execute the M-service locally.
    9: retrieve all required data from broadcast
       channels.
   10: continue M-service execution at the wireless
       device.
   11: repeat steps 6 through 10 for each service
       required.
   12: return results to the user.
End
```

6.1.3 Broadcast Content

Let us now discuss the information that needs to be provided by a broadcast based M-services system to its users.

Wireless Data

Wireless data is what mobile users ask for to satisfy their needs. In this book, wireless data is considered to be real-time information, such as traffic report, weather forecast, event schedule, availability for movie tickets. They are made available to mobile users through broadcast channels and may be updated frequently.

In the proposed infrastructure, wireless data consists of different *data sets*. A data set is similar to a table in relational databases. It has a name and a fixed schema which defines the structure of the data set. Each data set serves a specific purpose. For example, a "Traffic Report" data set would contain live traffic information. An M-service may need to access one or more data sets. Each service not only knows what data sets to access but also knows the structures of these data sets in order to filter on specific fields.

M-services

M-services are executables that provide various services to mobile users. They are made available through broadcast channels and executed on users' wireless devices, such as laptops and palms. In the proposed infrastructure, each service is assigned a unique service key which can be used to identify the service. When executed, a service may require input from an end user or another service. A service may also need to retrieve data from wireless channels to satisfy users' requests.

Assume a "Traffic Report" service generates live traffic reports for the area of the given zip code. Once the service is downloaded and invoked by a user. It would ask the user to enter a zip code. Once the zip code is entered, the service would then access broadcast channels to retrieve traffic information for the specified zip code and display it to the user.

Service Descriptions

The *Web Services Description Language (WSDL)* is an XML-based language that provides a model for describing Web services. The description contains information on how to communicate using web services. A client program connecting to a Web service can read the WSDL to determine what functions are available. Any special data types used are embedded in the WSDL file in the form of XML Schema. The client can then use SOAP to actually call one of the functions listed in the WSDL. The following example shows the WSDL definition for the "Traffic Report" service mentioned earlier.

```
<?xml version="1.0"?>
<definitions name="TrafficReport"
  targetNamespace="traffic"
  xmlns="http://schemas.xmlsoap.org/wsdl/"
  xmlns:soap="http://schemas.xmlsoap.org/wsdl/soap/"
  xmlns:tns="traffic" xmlns:plnk=
  "http://schemas.xmlsoap.org/ws/2003/05/partner-link/"
  xmlns:xsd="http://www.w3.org/2001/XMLSchema">
  <message name="TrafficReportRequestMsg">
    <part name="zipCode" type="xsd:string"/>
  </message>
  <message name="TrafficReportResponseMsg">
    <part name="report" type="xsd:string"/>
  </message>
  <portType name="TrafficReportPortType">
    <operation name="getReportByZipCode">
      <input name="getReportByZipCodeInput"
          message="tns:TrafficReportRequestMsg"/>
      <output name="getReportByZipCodeOutput"
```

```
                 message="tns:TrafficReportResponseMsg"/>
    </operation>
  </portType>
  <binding name="TrafficReportBinding"
           type="tns:TrafficReportPortType">
    <soap:binding style="document" transport=
       "http://schemas.xmlsoap.org/soap/http"/>
    <operation name="getReportByZipCode">
      <soap:operation soapAction=""/>
      <input name="TrafficReportInput">
        <soap:body use="literal"/>
      </input>
      <output name="TrafficReportOutput">
        <soap:body use="literal"/>
      </output>
    </operation>
  </binding>
  <service name="TrafficReportService">
    <port name="TrafficReportPort"
          binding="tns:TrafficReportBinding">
      <soap:address location=
         "http://server.com/TrafficReport1"/>
    </port>
  </service>
  <plnk:partnerLinkType name="TrafficReportService">
    <plnk:role name="TrafficReportServiceProvider">
      <plnk:portType
          name="tns:TrafficReportPortType"/>
    </plnk:role>
  </plnk:partnerLinkType>
</definitions>
```

As can be seen from the example, the WSDL record defines the input/output parameter types, the supported operation and the service location. The service location is defined by attribute location of soap:address element under service. For a regular Web service, the location would be an URL that the service can be accessed at. In the proposed infrastructure, all services are available on broadcast channels and can only be executed locally. In this case, the location would be the information mobile users need to find the requested service in the broadcast channels. Since a service can be uniquely identified by its service key. By default, we assume the location attribute stores the service key of the requested service.

Composite Services

As already mentioned, service composition is an important feature for Web services systems. It facilitates the definition of different business processes that involve multiple services. Business Process Execution Language (BPEL) for Web services is an XML-based language designed to enable task-sharing for a distributed computing or grid computing environment using a combination of Web services. We use BPEL to define composite services in the proposed infrastructure. The following example

shows how to use BPEL to define a simple process that invokes the "Traffic Report"
service.

```
\begin{verbatim}
<process name="TrafficReportProcess"
    targetNamespace="traffic"
    xmlns:tns="traffic" xmlns=
    "http://schemas.xmlsoap.org/ws/2003/03/business-process/">
  <partnerLinks>
    <partnerLink name="caller"
        partnerLinkType="tns:TrafficReportService"
        myRole="TrafficReportServiceProvider"/>
  </partnerLinks>
  <variables>
    <variable name="request" messageType=
        "tns:TrafficReportRequestMsg"/>
    <variable name="response" messageType=
        "tns:TrafficReportResponseMsg"/>
  </variables>
  <sequence>
    <receive name="TrafficReportReceive"
        partnerLink="caller"
        portType="tns:TrafficReportPortType"
        operation="getReportByZipCode"
        variable="request"
        createInstance="yes"/>
    <reply name="TrafficReportReply"
        partnerLink="caller"
        portType="tns:TrafficReportPortType"
        operation="getReportByZipCode"
        variable="response"/>
  </sequence>
</process>
```

Service Registry Information

In order for mobile users to discover suitable services, a registry that contains information about all services needs to be provided by the proposed infrastructure. UDDI
(Universal Description, Discovery, and Integration) [61] is the well known standard
used for Web services publication and discovery. The UDDI standard defines a rich
set of publishing and inquiry API functions for service providers to publish and man-
ager their services in the registry and for consumers to discover services that meet
their requirements. In a broadcast-based M-services system, a registry that contains
information about all services is also needed in order for mobile users to discover
suitable services. However, the UDDI standard is designed for interaction oriented
systems. It often takes several iterations before suitable services are found using the
defined inquiry API functions. This is obviously not suitable for broadcast-based
environment. Furthermore, it would be too expensive and inefficient to broadcast
a full-fledged UDDI registry. As already mentioned, the proposed infrastructure is
intended for information oriented Web services in a broadcast-based wireless envi-

ronment. Mobile users only need descriptive information about provided services in order to find a suitable service. We propose a simple registry service that provides information about services to mobile users. We define the following schema for each record stored in the registry.

```xml
<?xml version="1.0" encoding="UTF-8"?>
<xs:schema targetNamespace="mservice-registry"
           xmlns="http://www.w3.org/2001/XMLSchema"
           xmlns:xsd="http://www.w3.org/2001/XMLSchema"
           xmlns:tns="mservice-registry"
           elementFormDefault="qualified" version="0.1">
  <xsd:element name="service" type="tns:service_type"/>
  <xsd:complexType name="service_type">
    <xsd:sequence>
  <xsd:element name="name" type="xsd:NCName"/>
      <xsd:element name="description" type="xsd:string"
                   minOccurs="0"/>
      <xsd:element name="type">
        <xsd:simpleType>
          <xsd:restriction base="xsd:string">
            <xsd:enumeration value="simple"/>
            <xsd:enumeration value="composite"/>
          </xsd:restriction>
        </xsd:simpleType>
      </xsd:element>
      <xsd:element name="modelKey" type="xsd:integer"/>
      <xsd:element name="modelLocation" type="xsd:string"/>
    </xsd:sequence>
  </xsd:complexType>
</xsd:schema>
```

As can be seen from the schema, a registry record would contain a required *name* element which is used to identify a provided service. There is an optional description element which is intended to store information about the service. A required *type* element is used to determine the type of the service. It is an enumeration data type which has two possible values, *simple* and *composite*. The schema also defines a *modelKey* element which is used to reference the associated service description, i.e. WSDL or BPEL record. In the proposed infrastructure, we assume each WSDL or BPEL record is assigned a unique number. The *modelKey* element contains the number of the WSDL or BPEL record for the service. If the type is simple, the *modelKey* would reference the WSDL record for the service. Otherwise, it would reference the associated BPEL record. The *modelLocation* indicates where to find the WSDL or BPEL record. In this case, it would be the information of the broadcast channel that carries the WSDL or BPEL record. Shown below is the registry record for the "Traffic Report" service.

```xml
<service>
  <name>TrafficReportService</name>
  <description>This service provides traffic information
      based on given zip code.
  </description>
```

```
    <type>composite</type>
    <modelKey>1005</modelKey>
    <modelLocation>channel-23</modelKey>
</service>
```

6.1.4 Access M-services and Wireless Data

An M-services system deliver service registry, WSDL and BPEL records, M-services, and data records to mobile users through broadcast channels. Using the broadcast information, mobile users can discover and access services of their interest and obtain required wireless data by invoking operations provided by these services. The following procedure shows the steps for accessing a service.

```
Procedure access_service

Begin
    1: Retrieve service registry
    2: A service S is selected by user
    3: Retrieve WSDL records
    4: if (S is a simple service)
    5:    Load the WSDL record for service S
    6:    An operation is selected by the user
    7:    Required input is entered by the user
    8:    The selected service is downloaded and
             executed
    9:    The selected operation is invoked
   10:    Required data records are retrieved
   11: else if (S is a composite service)
   12:    Retrieve BPEL records
   13:    Load the BPEL record for service S
   14:    Load WSDL records for all required services
   15:    Required input is entered by the user
   16:    Required services are downloaded and executed
   17:    Required data records are retrieved
   18: end if
End
```

As can be seen from the above procedure, a mobile user first uses service registry to selected a service of his/her interest. If the selected service is a simple service, the user can then select an appropriate operation and enter required parameters based on its WSDL record. The service is then downloaded and executed. At last, requested data records are retrieved and presented to the user. If the selected service is a composite service, then the same process is repeated for all services defined in its BPEL record based on their dependencies.

Let us now use two example scenarios to illustrate how mobile users could to use the broadcast based M-services to satisfy their needs. Assume professor Edward has come to New York city to attend an international conference. The professor is new to this city. He is equipped with a wireless device that has access to a local M-services system that provides various services to tourists. In his stay in New York city, he would like to use his wireless device to make his trip more convenient. Upon arriving at New York city, professor Edward receives a welcome message from the local M-services system. The message includes the information required to access the system.

Scenario 1: accessing a single service

Let us consider a simple scenario first. After enjoying different attractions in the city, professor Edward would like to go to a French restaurant close to where he is. For that purpose, he uses his wireless device to access the local M-services system. At first, the wireless device retrieves registry, WSDL and BPEL records for all provided services. The professor then looks through these entries for a suitable service and notices a service `restaurant_finder`, which defines an operation for searching for restaurants based on their specialties (e.g., French) and addresses. Professor Edward selects this service. As indicated by the registry record, the `restaurant_finder` service is a simple service. The mobile device then retrieves the service from the system. After the service is downloaded, the professor selects to invoke an appropriate operation provided by this service. The professor is then prompted for required inputs and selects "French food" as the specialty and enters the name of the street he is currently on. The service then instructs the mobile device to retrieve the required data records. In a little while, the professor is presented with a list of restaurants meeting the specified requirements. He finally makes up his mind and selects a French restaurant on the same street based on the retrieved information.

Scenario 2: accessing a composite service

Let us now consider a more complicated scenario. Assume professor Edward just finished his presentation at the hotel where the conference is held. He would like to catch a Broadway show which starts in thirty minutes. He wants to know the fastest approach to get to the theater. The professor looks through the registry records and finds a composite service `transportation_finder`. According to the description of this service, it is used to find the best transportation between two given locations based on the routes and traffic conditions. According to its BPEL record, this composite service requires five services, namely *Bus_info*, *Subway_info*, *Taxi_info*, *Traffic*, and *Direction*. Figure 6.3 illustrates the invocation sequence defined by this composite service.

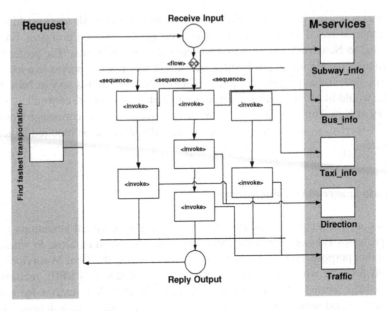

Fig. 6.3 Composite Service Example

Professor Edward selects the service and the mobile device follows the same process as stated in Scenario 1 to retrieve each required service from the system. Services *Bus_info*, *Subway_info* and *Taxi_info* are executed as soon as they are downloaded. They provide information on the routes between the two given locations (hotel and theater) for different transportation means. The *Traffic* service is invoked after *Bus_info* and *Taxi_info* for traffic considerations. The *Direction* service is invoked after *Bus_info* and *Subway_info* to calculate the walking distance between the locations and the their closest bus stops or subway stations. At last, all possible routes together with estimated time and costs are presented to professor Edward, who can then make his decision based on the provided choices.

6.2 Semantic Access to Composite M-services

Access efficiency is one of the most important issues in wireless broadcast environments. The efficiency is usually measured by response time and power consumption. Mobile users always want to obtain requested information with the least possible time and power consumption. Data access methods are used in wireless data broadcast systems to help users retrieve wireless data efficiently. A data access method defines the way wireless data records are organized in broadcast channels and the protocol for clients to access them. Accessing information in an M-services system, on the other hand, is different from accessing data in a wireless data broadcast system. The access efficiency is affected by different semantics. For example, a composite service may invoke multiple services. Based on the BPEL definition, these services could have certain dependencies between each other. This means services can only be executed conforming to the defined dependencies. Therefore, it is important to know the type of a service and its defined dependencies to access and execute the service. Furthermore, services and data records are delivered on "push" instead of "pull" basis. This means some services may arrive when they are not ready to be executed yet. On some mobile devices, based on available resources, they could be cached to be executed later. Alternatively, they would have to be downloaded on their next arrivals. In this chapter, we discuss how to access M-services efficiently based on these semantics.

6.2.1 Semantics for M-services Systems

First, let us define the access semantics for broadcast based M-services. There are two types of semantics, namely *service semantics* and *client semantics*. Service semantics define characteristics of services that may have impact on access efficiency. The following two service semantics are considered in this book:

- *Service dependencies*: Service dependencies define the access sequence of a composite service. This information can be obtained from the BPEL record of a composite service. Services must be retrieved and executed according to the defined service dependencies. This information could be used to come up with better access strategies.
- *Service locations*: Service locations indicate the exact location of each service in broadcast channels. When accessing multiple services at the same time, this information could be used to access these services more efficiently.

Different mobile devices have different system configurations and capabilities. Some models are faster than the others. These differences result in different access efficiencies. Client semantics define capabilities of mobile devices. The following semantics are considered in this book:

- *Number of receivers (n_r)*: A mobile device may be equipped with more than one receivers. Each receiver can independently listen to broadcast channels and retrieve information. In a multi-channel broadcast environment, mobile devices equipped with multiple receivers could listen to multiple broadcast channels concurrently and therefore could have much better access efficiency.
- *Maximum number of services that can be executed concurrently (n_e)*: This factor usually reflects the computation power of a mobile device. A BPEL record may define multiple services that are independent of each other. In this case, allowing them be executed at same time could be a huge advantage.
- *Maximum number of services that can be cached (n_c)*: Sometimes a receiver may come across a required service which is dependent on the results of another

service, which has not been retrieved yet. A mobile device may have to wait till the next broadcast cycle for the same service to arrive again. In this case, the service can be retrieved and cached until the depended service is executed. However, there could be different limitations on how many services could be cached based on different types of mobile devices.

6.2.2 Access Action Tree

Service dependencies semantic defines access sequences of composite services. We propose to use an *access action tree* to capture all service dependencies in a BPEL record. An access action tree defines all actions for accessing services and data records from broadcast channels in the process of executing a composite service. There are two types of nodes in an access action tree, as shown in Table 6.1.

Table 6.1 Access action tree nodes

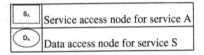

As can be seen from the table, an action for downloading a service is represented by a *service access node* and an action for retrieving all data records required by a service is represented by a *data access node*. In an access action tree, a BPEL *<sequence>* element is represented by parent-child relationships and a BPEL *<flow>* element is represented by sibling relationships. Figure 6.4 shows the access action tree for composite service `transportation_finder` mentioned earlier. Subscripts *BI, SI, TI, T,* and *D,* represent services *Bus_info, Subway_info, Taxi_info, Traffic,* and *Direction* respectively.

6.2.3 Traversing Algorithms

An access action tree captures all actions for retrieving services and wireless data, and their dependencies. Let us now discuss how mobile devices can use access action trees to determine the sequence of service and data retrievals. In this section, we present three algorithms for traversing access action trees.

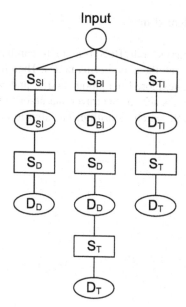

Fig. 6.4 Access Action Tree Example

6.2.3.1 Depth-first Algorithm

The first algorithm, *depth-first-access*, is based on the well-known depth-first search (DFS) algorithm for traversing trees and graphs [10]. This algorithm starts at the root and explores as far as possible along each branch before backtracking. Using this algorithm, action nodes in an access action tree are accessed from top to bottom which reflects their dependencies. The algorithm is shown as follows:

```
depth-first-access(tree) {
    access_sequence = () // empty list
    root = tree.root // get root node
    depth-first-traverse(access_sequence, root)
}

// recursive traverse
depth-first-traverse(access_sequence, node) {
    append node to access_sequence
    for each child i of node {
        depth-first-traverse(access_sequence, i)
    }
}
```

The algorithm generates an access sequence based on the traversing paths. The dependencies of the action nodes are guaranteed by the generated access sequence. This means when a service is executed, it is guaranteed that its depended services have already been completed. Applying this algorithm to the `transportation_finder` service would generate access sequence S_{SI}, D_{SI}, S_D, D_D, S_{BI}, D_{BI}, S_D, D_D, S_T, D_T, S_{TI}, D_{TI}, S_T, D_T.

6.2.3.2 Breadth-first Algorithm

In graph theory, breadth-first search (BFS) is a graph search algorithm that begins at the root node and explores all the neighboring nodes. Then for each of those nearest nodes, it explores their unexplored neighbour nodes, and so on, until it finds the goal [11]. The second access action tree traversing algorithm, *breadth-first-access*, is based on the breadth-first search algorithm. The algorithm is shown as follows:

```
breadth-first-access(tree) {
    access_sequence = () // empty list
    root = tree.root // get root node
    breadth-first-traverse(access_sequence, root)
}

// recursive traverse
breadth-first-traverse(access_sequence, node) {
    for each child i of node {
        append i to access_sequence
    }
    for each child i of node {
        depth-first-traverse(access_sequence, i)
    }
}
```

The algorithm generates an access sequence based on the tree levels. The higher level nodes always precede lower level nodes. This algorithm makes it possible for all parallel actions (sibling nodes) to be executed at the same time. Applying this algorithm to the `transportation_finder` service would generate access sequence S_{SI}, S_{BI}, S_{TI}, D_{SI}, D_{BI}, D_{TI}, S_D, S_D, S_T, D_D, D_D, D_T, S_T, D_T.

6.2.3.3 Closest-first Search

The *depth-first-access* and *breadth-first-access* algorithms generate access sequences completely based on access action tree structures. They guarantee all action nodes are executed once an access action tree is traversed. However, the generated access sequences may not be efficient because the tune-in positions of mobile devices are random. At any point, the next service in an access sequence could be far away from the current tune-in position. For example, according to the access sequence generated for the `transportation_finder` service by the *depth-first-access* algorithm, S_{SI} should be accessed and executed before S_{BI}. However, if S_{BI} arrives before S_{SI} when a mobile device tunes into a broadcast channel, it would take less time to download S_{BI} first and then S_{SI}. Therefore, it would be more efficient to try to retrieve services and wireless data based on the distances from the current tune-in position to their broadcast positions. For this reason, we propose a *closest-first-access* algorithm that takes into consideration service and data positions. Instead of generating a static access sequence, the algorithm maintains a dynamic access sequence. The access sequence is updated according to the current tune-in position of a mobile device.

```
// init access sequence
closest-first-access(tree, initial-tune-in_pos) {
```

```
    access_sequence = () // empty list
    root = tree.root // get root node
    depth-first-traverse(access_sequence, root)
    // or breadth-first-traverse(access_sequence, root)
    closest-first-access-sort(access_sequence, initial_tune-in_pos)
}

// sort access sequence based on distances from tune-in position
// tune-in_pos: current tune-in position
closest-first-access-sort(access_sequence, tune-in_pos) {
    n = length of access_sequence
    for (i=0; i<n; i++) {
        c = access_sequence[i]
        d = distance from tune-in_pos to c's broadcast position
        for (j=0; j<i; j++) {
            c1 = access_sequence[j]
            d1 = distance from tune-in_pos to c1's broadcast position
            if (c < c1) {
                insert item i in front of j
                break
            }
        }
    }
}
```

The *depth-first-access* and *breadth-first-access* algorithms generate static access sequences. Mobile devices follow the generated access sequences to perform access actions. The algorithms only need to be invoked once at the beginning. The *closest-first-access* algorithm, on the other hand, maintains a dynamic access sequence based on the tune-in positions of a mobile device. It needs to be invoked every time after a data access node is completed. This is because the required data sets for a service and their positions in broadcast channels for a service are not known until the service is actually executed. The first access sequence is generated based on the initial tune-in position of a mobile device. It needs to be updated with a new tune-in position after each data access action.

6.2.4 Multi-channel Access Method

We have discussed two access methods for accessing composite services. The proposed methods are based on the assumption that a mobile device can only listen to one channel and execute one service at a time. With recent development in mobile wireless technologies, mobile devices are now capable of listening to multiple chan-, nels at the same time and execute a few applications concurrently. Furthermore, a service that is dependent on other services could be cached locally and executed later when all depended services are completed. In this section, we discuss access to composite services in a more generalized way. We propose a multi-channel method that takes these system factors into consideration. The method provides an efficient way to access services and wireless data based on these factors and the BPEL definition of a composite service.

6.2.4.1 Component View

The multi-channel access method defines three independent processes, namely, *service request process*, *receiver process*, and *service execution process*. Figure 6.5 shows the relationships of these processes.

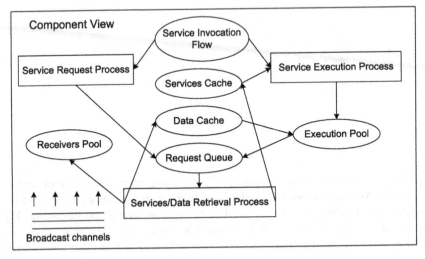

Fig. 6.5 Component View

The service request process is responsible for downloading required services defined in the access action tree of a composite service. This process generates service retrieval requests and store them in *request queue*. The receiver process picks up requests from the request queue and download services from broadcast channels. Once downloaded, services are stored in *services cache*. The *service execution process* obtains services from the services cache and execute them in the sequence as defined in the access action tree. When a service is executed, a data retrieval request is generated and stored in the request queue. The request defines data set names and filtering criteria for retrieving required data records. The receiver process will pick up and process the data retrieval requests. After all data records for a service are retrieved, they are stored in data cache. These data records will either be final results required by mobile users or data required by other services. The *receivers pool* contains all the receivers that are equipped by a mobile device. Each receiver can independently listen to broadcast channels and retrieve services or data records. The receivers pool is used by the receiver process to retrieve services and data records from broadcast channels. The *execution pool* loads and executes services. The size of the execution pool reflects the number of services that can be executed concurrently on a mobile device.

6.2.4.2 Process Procedures

In this section, we present the the procedure for each process defined in the multi-channel access method.

Service Request Process

The service request process is for generating service requests based on the given access action tree and the size of the services cache at any time. The key for this process is to determine whether a service should be retrieved when it arrives. We define the following condition for this purpose.

Condition 1: A service request can be inserted into the request queue if and only if one of the following statements is true:

- The service is not dependent on another service.
- All services this service is dependent on have been executed and their required data records have been retrieved.
- The total size of the services that this service is dependent on does not exceed the available space in the services cache.

Algorithm *service-request-process* listed below shows how the service request process works.

```
service-request-process(tree) {
    wait_period = 1 second
    N = number of required services
    i = 0
    while i < N {
        request_generated = false
        sequence = closest-first-access(tree)
        for each service action x in sequence
            cs = get current service cache size
            if x is root or x's parent service is
                    completed or size of (x's
                    pending parent services) < cs {
                req = generate service request
                insert req into the request queue
                request_generated = true
                i = i + 1
            }
        }
        if (!request_generated)
            wait (wait_period)
        end if
    }
}
```

The *service-request-process* puts service actions in the request queue based on their positions in the access action tree. A service action is only placed in the request queue if one of the conditions defined in **Condition 1** is true. This process would pause for a specified period of time if the request queue is full or there is no service action meeting the criteria. The process completes when all service actions are placed in the request queue.

Receiver Process

The receiver process retrieves services and wireless data from broadcast channels. This process uses a pool of receivers to access broadcast channels. The process assumes the locations of requested services and data sets in broadcast channels are known. Algorithm *receiver-process* illustrates how receiver process works.

```
receiver-process() {
    N = number of required services
    item_taken = false
    wait_period = 1 second
    while i < N {
        r = get an available receiver
        if r != null {
            r tunes into a broadcast channel
            p = get current tune-in position
            selected_item = get first item in
                request queue
            min = calculate distance between p and
                selected_item
            for each item k in request queue {
                l = get location of k
                d = calculate distance between
                    p and l
                if min == -1 or d < min {
                    min = d
                    selected_item = k
                }
            }
            r takes selected_item
            remove selected_item from request queue
            if selected_item is a data request {
                i = i + 1
            }
        } else {
            wait (wait_period)
        }
    }
}
```

```
        }

        // this method is invoked when an item is retrieved
        item-received(receiver, item) {
            mark receiver as available
            if item is a service request {
                store item in service cache
            else if item is a data request {
                store item in data cache
            }
        }
```

The *receiver-process* picks up requests from the request queue and retrieves requested items. A requested item could either be a service or a set of data records. An
active receiver would always try to pick up the closest item from its current tune-in
position. The process would pause for a specified period of time if all receivers are
busy. The process completes if data records for all requested services are retrieved.
When a service is retrieved, it is placed in the service cache waiting to be executed.
When a set of data records are retrieved, they are placed in the data cache to be used
by other services or presented to the user.

Service Execution Process

The service execution process takes retrieved services from the service cache, and
execute them in the sequence as defined by the access action tree. Upon its execution, each service knows which data sets to visit and what data records to retrieve.
The service execution process also generates data requests after a service is executed and saves them in the request queue. The key to this process is to determine
whether a service in the service cache should be executed. We define *Condition 2*
for this purpose.

Condition 2: A service in the services cache can be executed if and only if the
following two statements are both true:

- All depended services have been completed.
- The execution pool can still accept new services.

Algorithm *service-execution-process* illustrates how a mobile device would execute services.

```
service-execution-process()
    wait_period = 1 second
    N = number of required services
    i = 0
    while i < N {
        service_started = false
        for each item i in service cache {
```

```
            if i's parent is completed
                    and execution pool is not full {
                add i to execution pool
                start execution
                remove i from service cache
                service_started = true
                i = i + 1
            }
        }
        if (!service_started) {
            wait (wait_period)
        }
    }
}
// a service knows what data records to retrieve
// this method is invoked with generated data request
service-executed(data_request) {
    insert data_request into request queue
}
```

The *service-execution-process* executes services in the service cache based on
the capacity of the service execution pool. A service is executed if all its depended
services have been completed and the execution pool is not full. When a service is
executed, a data request is generated and placed in the request queue.

A Simple Example

Let us now use a simple example to illustrate how the multi-channel access method
works. Assume a mobile user wants to run a composite service, *CS1*, which has the
following BPEL definition:

```
<sequence name="main">
    <receive name="receiveInput" partnerLink="query"
            portType="query:Example"
            operation="process"
            variable="inputVariable"/>
    <sequence name="Sequence_1">
        <invoke name="InvokeA" partnerLink="A"/>
        <flow name="Flow_1">
            <sequence name="Sequence_2">
                <invoke name="InvokeB" partnerLink="B"/>
                <invoke name="InvokeD" partnerLink="D"/>
            </sequence>
            <invoke name="InvokeC" partnerLink="C"/>
        </flow>
```

```
</sequence>
<reply name="replyOutput" partnerLink="query"
        portType="query:Example" operation="process"
        variable="outputVariable"/>
</sequence>
```

As can be seen from the record. Four services, *A*, *B*, *C*, and *D*, are required. There are two sequential and two parallel service invocation sequences. We assume that there are four M-services channels. We also assume that the locations of these four services in the broadcast channels are as shown in Figure 6.6.

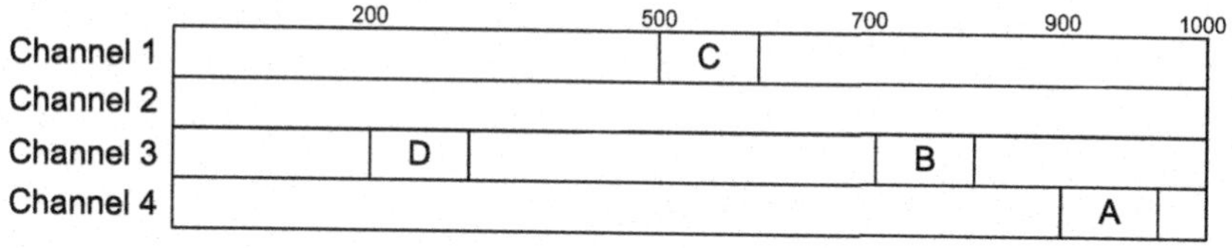

Fig. 6.6 A simple example

To simplify the scenario, we assume the mobile device has only one receiver and can execute only one service at a time. The procedure below illustrates how the multi-channel access method executes *CS1*. The procedure captures the steps for all processes.

```
Procedure a-simple-example
REQ : service request process
EXE : service execution process
RET : receiver process
Begin
    1. Retrieve BPEL record
    2. Load M-services physical index
    3. Load data index
    4. Tune into an M-services channel
    5. REQ: Generate service request for service
            A and insert it into the request queue
    6. RET: Retrieve service A from M-services channels
    7. RET: Save service A in the services cache
    8. EXE: Execute service A
    9. EXE: Generate data request for service A
            and insert it into the request queue
   10. RET: Retrieve data records for service A
   11. Repeat steps 7 to 12 for services B, C, and D.
End
```

Chapter 7
Broadcast Channel Organization

A broadcast channel organization determines how information is organized in broadcast channels. Mobile devices must know the channel organization in order to find required information in a broadcast channel. The actual channel organization has direct impact on access efficiency because it determines how the broadcast information can be retrieved. The most straightforward channel organization is called "flat broadcast", which is to directly broadcast data without providing any additional information on how to locate the broadcast data. With flat broadcast, mobile devices must stay active all the time to filter retrieved information in order to find requested data. This is obviously not efficient especially for mobile devices with limited power supply. In this chapter, we propose a few channel organizations suitable for the broadcast-based M-services infrastructure. For each channel organization, we also present access protocols for mobile devices to retrieve information from these channels.

In a broadcast based M-services system, a mobile device may need to access different broadcast channels, such as registry, BPEL, WSDL, M-services, and data channels, in order to satisfy a user request. As already discussed, for a new user to lookup, retrieve, and execute a service, the following steps are followed:

1. Download registry records
2. Select a service
3. Retrieve WSDL/BPEL records for the selected service
4. Enter input parameters if required
5. Retrieve the actual service(s)
6. Execute the services(s)
7. Retrieve required data records

Steps 1, 3, 5, and 7 require mobile devices to access broadcast channels. Since registry, BPEL and WSDL records could be cached, we assume the access and tuning times for steps 1 and 3 are constants. In our analysis, we focus on studying the efficiency of steps 5 through 7, which are for accessing and executing services and retrieving wireless data. We define the following parameters for measuring the efficiency of the proposed channel organizations:

X. Yang and A. Bouguettaya, *Access to Mobile Services*, Advances in Database Systems 38, DOI: 10.1007/978-0-387-88755-5_7, © Springer Science + Business Media, LLC 2009

- *Total access time (T_{at})*: This is the total time spent by all receivers of a mobile device to retrieve information from different broadcast channels in order to fulfill a user request. This parameter indicates how long the receivers are kept busy to finish a user request.
- *Total tuning time (T_{tt})*: This is the total tuning time spent by all receivers of a mobile device to retrieve information from different broadcast channels in order to fulfill a user request.

In this chapter, we discuss these channel organizations in details. We also provide analytical model for each proposed organization. Table 7.1 lists the symbols used in the analytical models.

Table 7.1 Symbols for composite M-services

S_m	Average service size
N_m	Average number of services per channel
S_{mk}	Service key size
S_d	Average data set size
N_d	Average number of data sets in a data channel
S_{dk}	Data set key size
D	Data rate of a broadcast channel
N_s	Number of services per BPEL (N_s=1 for simple services)
N_p	Number of required data sets per simple service

As already discussed, the efficiency of accessing composite service is affected by different semantics at runtime. It is difficult to derive analytical model for a channel organization considering all these semantics. To simplify the theoretical analysis, we assume mobile devices can only listen to one channel at a time and execute one service at a time. We also assume a service cannot be cached to be executed later. With these assumptions, we guarantee that all required services must be executed sequentially.

7.1 Flat Broadcast

Let us begin the discussion with the most basic way of broadcasting M-services and data records, *flat broadcasting*. With flat broadcasting, no particular information is provided in broadcast channels to help users locate services or data. Let us assume the WSDL record of each required service contains the channel number and the service key. A mobile device would need to scan through all indicated service channels to find required services. During this process, the mobile device must stay active at all time. Once the service is downloaded and executed. The mobile device then tunes into a data channel to retrieve all requested data records. It would take an average of half of the broadcast cycle in a service channel to find each required service. Assume data sets are randomly distributed in data channels. It would take an average of $\frac{(2N_p-1)}{2N_p}$ broadcast cycle of a data channel to retrieve all N_p required data records. Thus, the total access and tuning times can be expressed as follows:

$$T_{at} = T_{tt} = (\frac{1}{2}N_s S_m N_m + \frac{(2N_p-1)}{2N_p} S_d N_d)/D$$

7.2 Selective Tuning

The concept of *selective tuning* is first introduced in [26]. With selective tuning, mobile devices only selectively tune into broadcast channels when required. Selective tuning is usually achieved with extra information (e.g. index) in the broadcast channel to help determine when the requested data would arrive. Therefore, mobile devices do not have to actively listen to the channel all the time. This obviously preserves power consumption, which is crucial to mobile devices with limited power supply. We can also apply selective tuning to accessing services and data channels. Instead of staying active all the time, mobile devices can selectively read the service keys and data set keys to match with the requested service keys and data set keys. Only when there is a match, a mobile device would proceed to read the service or data set that follows the matched key. This would keep the mobile device in doze mode most of the time which reduces the total tuning time and results in better power consumption. Since different services and data sets are in different sizes, mobile devices need to know the size of each service and data set to decide how long to stay in doze mode each time. We add a size field immediately following the key field for this purpose. Figure 7.1 shows the service channel structure for selective tuning. Data channels have the same structure as service channels.

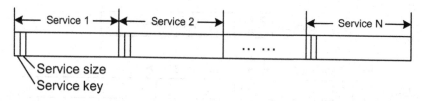

Fig. 7.1 Service Channel with Selective Tuning

Let S_p be the length of the size field. Let S_{ma} be $S_m + S_{mk} + S_p$ and S_{da} be $S_d + S_{dk} + S_p$. The total access and tuning times can be expressed as follows:

$$T_{at} = N_s(\frac{1}{2}S_{ma}N_m + \frac{(2N_p-1)}{2N_p} S_{da}N_d)/D$$

$$T_{tt} = N_s(\frac{1}{2}(S_{mk}+S_p)N_m+S_mN_s+\frac{(2N_p-1)}{2N_p}(S_{dk}+S_p)N_d+S_dN_p)/D$$

7.3 Indexed Broadcast

The selective tuning broadcast drastically reduces the total tuning time, which means better power consumption. However, mobile devices still need to read the key field for each service or data set. In this section, we propose three *indexed broadcast* methods, which can lead mobile devices to the exact locations of the requested services or data sets. With indexed broadcast, each channel has a channel number which can be used to identify the channel. The smallest logical unit of each channel is called a *bucket*. Each bucket is of the same size in all channels. Broadcast information such as services or data items span at least one bucket. Every bucket has a sequence number which indicates its position in a broadcast channel. A piece of information can be located using the combination of the channel number and bucket sequence number. The proposed indexed broadcast methods use index to help mobile users to locate required services and data sets. Each index contains the channel number and bucket position of the corresponding service or data set. Figure 7.2 shows the layout of M-services and data channels for indexed broadcast.

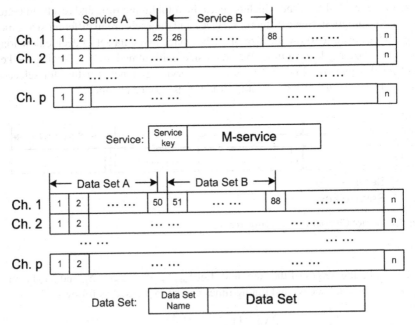

Fig. 7.2 M-services and Data Channels for Indexed Broadcast

An M-service is placed in a number of consecutive buckets. The position of an M-service can be determined by its channel number and the starting bucket number. As can be seen from the figure, service A starts at bucket 1 and ends at bucket 25 on channel 1. Service B can be found at bucket 26. Similarly, the location of a data set is determined by the channel number and its starting bucket number. When a service is executed, it knows the names of the required data sets. The mobile device would listen to data channels to look for the required data sets and then filter for the requested records.

7.3.1 Predefined Index

With *predefined index*, the index of each service is stored in its WSDL record. Each service knows the index of the required data sets. A mobile device can use the index information to locate the required services and data sets in broadcast channels. To access a simple service, a mobile device takes the following steps (assuming the WSDL record has already been retrieved):

1. Read the index of the requested service from the WSDL record
2. Obtain the channel number and bucket position from the index
3. Tune into the broadcast channel designated by the channel number
4. Read the next arriving bucket and obtain the bucket number
5. Calculate the time of arrival based on the bucket numbers of the current bucket and the bucket position of the requested service
6. Go into doze mode
7. Wake up at the estimated time of arrival
8. Read the next bucket and make sure the service key matches with that of the required service
9. Download the service
10. Execute the service
11. Retrieve required data records based on the data set index provided by the service.

Let S_b be the size of a bucket. The total access and tuning times can be derived as follows:

$$T_{at} = N_s(\frac{1}{2}S_m N_m + \frac{(2N_p - 1)}{2N_p}S_d N_d)/D$$

$$T_{tt} = N_s(S_b + (\lceil\frac{S_m}{S_b}\rceil + N_p\lceil\frac{S_d}{S_b}\rceil)S_b)/D$$

7.3.2 Index Channel

The predefined index method further improves the total tuning by allowing mobile devices to stay in doze mode most of the time. However, this method has the following limitations:

- The positions of services are tightly coupled with WSDL records.
- A WSDL record can only be used by one service because it contains the service location information.
- The positions of data sets are tightly coupled with services.
- WSDL records need to be updated if service positions change.
- Services need to be updated if the positions of the required data sets change.

We now propose another variation of the indexed broadcast, *index channel* method. With this method, two separate index channels are introduced, *M-services index channel* and *data index channel*. The M-services index channel is used to help users locate and retrieve M-services. The M-services index channel contain information about the locations of all M-services in the M-services channels. Users access the M-services index channel first to find the locations of the requested services. Then they tune into M-services channels based on the obtained location information to retrieve the requested services. Figure 7.3 shows the structure of the M-services index channel.

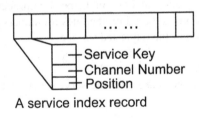

A service index record

Fig. 7.3 M-services Index Channel

The index channel contains physical indices that tell mobile users where different services are in M-services channels. Each index is in the form of a triplet *(k, c, p)*, where *k* is the service key, *c* is the channel number and *p* is the sequence number of the starting bucket for an M-service. Physical index is small in size. For example, a typical service key is usually 16 bytes. Let us assume a bucket sequence number takes 4 bytes (the size of an integer in Java). Then the total index size for 1000 M-services is only 20K bytes, which can be easily delivered to mobile users in less than a second. The physical index can be saved on users' mobile devices for reuse. Users only need to retrieve the physical index when it is changed. A version number is included in the index channel. Any change between broadcast cycles would result in a new version number. The version number is also included in M-services channels. Preceding each service, there is not only its service key but also the version number for the corresponding index.

Similarly, a data index channel is used to help mobile devices locate the positions of the data sets. Figure 7.4 shows the structure of the data index channel.

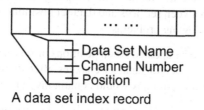

A data set index record

Fig. 7.4 Data Index Channel

As can be seen from the figure, the data index channel has the same structure as the services index channel. The data index channel consist of indices for locating all broadcast data sets. Each index is in the form of a triplet (d, c, p), where k is the data set name, c is the channel number and p is the sequence number of the starting bucket for the data set. The content of the data index channel could also be cached on mobile devices for reuse. A version number is also used in the data index channel and data channels to indicate changes between broadcast cycles. Let S_v be the size of the version number, S_{mc} be $S_m + S_{mk} + S_v$, and S_{dc} be $S_d + S_{dk} + S_v$. The total access and tuning times can be expressed as follows:

$$T_{at} = (S_o + N_s(\frac{1}{2}S_{mc}N_m + \frac{(2N_p - 1)}{2N_p}S_{dc}N_d))/D$$

$$T_{tt} = N_s(S_o + S_b + (\lceil\frac{S_{mc}}{S_b}\rceil + N_p\lceil\frac{S_{dc}}{S_b}\rceil)S_b)/D$$

where S_o is the overhead introduced by retrieving up-to-date services and data index. Let S_{mi} be the broadcast cycle length of the services index channel and S_{di} be the broadcast cycle length of the data index channel. We have

$$S_o = \begin{cases} 0 & \text{if } v_{mb} = v_{mc}, v_{db} = v_{dc} \\ S_{mi} & \text{if } v_{mb} \neq v_{mc}, v_{db} = v_{dc} \\ S_{di} & \text{if } v_{mb} = v_{mc}, v_{db} \neq v_{dc} \\ S_{mi} + S_{di} & \text{otherwise} \end{cases}$$

where v_{mb} and v_{mc} are the current and cached version numbers for the services index, v_{db} and v_{dc} are the current and cached version numbers for the data index respectively.

7.3.3 Interleaved Index

The index channel method avoids the tight coupling between WSDL records and service positions, and between services and data set positions. However, the method still suffers from the following problems:

- Cached index needs to be updated when positions of services or data sets change.
- If the position of a service or data set changes after its index is read from the index channel, the obtained index would contain invalid position information.

To guarantee accurate position information by index, we proposed the third variation of the indexed broadcast, *interleaved index*. With this method, index is interleaved with services or data sets. Figure 7.5 shows the structure of an M-services or data channel.

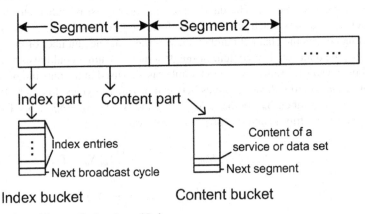

Fig. 7.5 Channel Layout for Interleaved Index

Each channel is divided into *p* partitions. Each partition consists of two parts, *index part* and *content part*. The content part of a segment contains the actual services or data sets allocated for that segment. The index part contains index for all services or data sets in that segment as well as all subsequent segments in the same broadcast cycle. For example, as shown in the figure, the index part of segment 1 contains index for services/data sets in all segments in a broadcast cycle. The index part of segment 2 contains index for all segments but segment 1. In each services channel, services are placed in the order of service keys. Similarly, in each data channel, data sets are placed in the order of data set names. Buckets in an index part are called *index buckets*. Buckets in a content part are called *content buckets*. An index bucket contains a few index entries and an offset to the next broadcast cycle. At the end of each content bucket, there is an offset field pointing to the next segment. The following procedure shows how to access a service in a broadcast channel:

```
Procedure access_service_interleaved_broadcast(key)
```

```
Begin
    1: tune into a broadcast channel
    2: read current bucket
    3: if (index bucket)
    4:    read index
    5:    current_key = first key in index
    6:    if (key < current_key)
    7:      wait till next broadcast cycle
    8:    else
    9:      keep reading until key is found
   10:      obtain service location
   11:      wait till the requested service arrives
   12:      retrieve the service
   13:    end if
   14:  else if (content bucket)
   15:      wait till the next segment
   16:      go to step 2
   17:  end if
End
```

With interleaved index, each broadcast cycle of a service channel is divided into p partitions. Within each partition, there would be up to $\lceil \frac{N_m}{p} \rceil$ services. Let S_i be the length of an index entry and S_f be the length of an offset value. It would take $\frac{\lceil \frac{N_m}{p} \rceil}{\lfloor \frac{S_b}{S_i} \rfloor}$ buckets to store $\lceil \frac{N_m}{p} \rceil$ index entries. Most index entries are repeated in multiple partitions. For example, index entries pointing to the services in the last partition are placed in all partitions. The total number of index buckets in all partitions for a service channel, N_{si}, can be calculated as follows:

$$N_{si} = \frac{\lceil \frac{N_m}{p} \rceil}{\lfloor \frac{S_b}{S_i} \rfloor}(1+2+...+p)$$

$$= \frac{\lceil \frac{N_m}{p} \rceil}{\lfloor \frac{S_b}{S_i} \rfloor}(\frac{p(p+1)}{2})$$

Let S_{MB} be the size of a broadcast cycle for a service channel. It is intuitive that $S_{MB} = N_{si}S_b + (S_m + S_{mk})N_m)$. Let S_{sp} be the average size of a partition which would be $\lceil \frac{S_{MB}}{p} \rceil$. A mobile device first needs to reach the partition that contains the index entry that points to the requested the service. The index entry can be found at the beginning of any partitions that precede the actual service. The time to reach that point is called *initial probe*. Let us assume the initial tune-in position is within the ith $(1 \leq i \leq p)$ partition. The requested service could either be in the $p - i$ partitions after the tune-in position or the p partitions preceding the tune-in position. The initial probe distance S_{ip} with tune-in position i can be represented as follows:

$$S_{ip} = \begin{cases} \frac{p-i}{p}\frac{S_{sp}}{2} & \text{if } i < x \le p \\ \frac{i}{p}(\frac{1}{2}+p-i)S_{sp} & \text{if } 1 \le x \le i \end{cases}$$

After reaching the partition that contains request index entry, the average distance to the actual service S_{sp} (*service probe distance*) can be calculated as follows:

$$S_{sp} = \begin{cases} \frac{p}{2}S_{sp} & \text{if } i < x \le p \\ \frac{p-i}{2}S_{sp} & \text{if } 1 \le x \le i \end{cases}$$

The access time to retrieve a service would be the sum of the initial probe time and the service probe time. The probability of the tune-in position falling in any partition i is $\frac{1}{p}$. The total service access time T_{sa} can be derived as follows:

$$
\begin{aligned}
T_{sa} &= \sum_{i=1}^{p} \frac{1}{p}(S_{ip} + S_{sp})/D \\
&= \frac{1}{p}S_{sp}\sum_{i=1}^{p}((\frac{p-i}{p}\frac{1}{2}+\frac{p}{2})+(\frac{i}{p}(\frac{1}{2}+p-i)+\frac{p-i}{2})/D \\
&= \frac{1}{p}S_{sp}(\frac{11}{12}p^2 + \frac{1}{4}p - \frac{1}{6})/D \\
&= S_{sp}(\frac{11p}{12} + \frac{1}{4} - \frac{1}{6p})/D
\end{aligned}
$$

Let q be the number of partitions in a data channel. The number of total index buckets in a data channel, N_{di}, can be derived similarly as N_{si}. We have $N_{di} = \frac{\lceil \frac{N_d}{q} \rceil}{\lfloor \frac{S_b}{S_i} \rfloor}(\frac{q(q+1)}{2})$. The broadcast cycle of a data channel is $S_{DB} = N_{di}S_b + (S_d + S_{dk})N_d)$. The average access time for retrieving N_p data sets would be $T_{da} = \frac{(2N_p-1)}{2N_p}SDB$. The total access time would be the sum of the service and data access times. As to the tuning time, the mobile device needs to read the first bucket after tune-in to determine the current position and the distance to the partition that contains requested index entry. Then it reads the index buckets until the requested index entry is found. It tunes in again when the request service or data set arrives and starts downloading the requested information. Therefore, the total access and tuning times for retrieving N_s services and their required data can be expressed as follows:

$$
\begin{aligned}
T_{at} &= N_s(T_{sa} + T_{da}) \\
T_{tt} &= N_s(2 + \frac{1}{2}\lceil\frac{N_{si}}{p}\rceil + \frac{(2N_p-1)}{2N_p}\lceil\frac{N_{di}}{q}\rceil + \lceil\frac{S_{ma}}{S_b}\rceil + \lceil\frac{S_{da}}{S_b}\rceil)S_b/D
\end{aligned}
$$

where $S_{ma} = S_m + S_{mk}$, $S_{da} = S_d + S_{dk}$ and N_{di} is the total number of index buckets in data channel.

7.3.4 Comparison

We have proposed three different index broadcast methods. The predefined index method has the best performance because it has the least overhead on services and data channels. However, as already mentioned, this method has a few limitations. WSDL records and service positions as well as services and data set positions are tightly coupled. It should not be used if services and data sets are updated frequently. The index channel method has slightly more overhead than the predefined method. But it avoids the coupling between WSDL records and service positions, and between services and data set positions. However, this method still suffers from frequent updates. The third method, interleaved index, has the most overhead. But it guarantees the accuracy of service/data set positions by broadcast index interleavely with services or data sets. This method should be used if frequent updates to services and data sets are expected.

Chapter 8
Implementation and Practical Study

In this chapter, we provide practical study of the proposed infrastructure, access methods, and channel organizations. A testbed is implemented for simulating accessing composite services in a broadcast-based environment. First, we present the architecture of the testbed. Then we discuss the experiments that were conducted for studying the efficiency of the proposed access methods and channel organizations. In these experiments, we also study the impact of different semantics on the access efficiency.

8.1 Implementation

An event-driven testbed has been implemented to simulate accessing composite services in a broadcast-based environment. The testbed is implemented in Java language. Figure 8.1 shows the main components of the testbed.

The testbed consists of two main components, *broadcast server* and *mobile clients*.

8.1.1 Broadcast Server

The broadcast server is responsible for creating different types of broadcast channels and simulating the broadcasting process. The following broadcast channels are supported:

- One WSDL channel
- One BPEL channel
- One or more services channels
- One services index channel (only required by index channel method)
- One or more data channels

X. Yang and A. Bouguettaya, *Access to Mobile Services*, Advances in Database Systems 38,
DOI: 10.1007/978-0-387-88755-5_8, © Springer Science + Business Media, LLC 2009

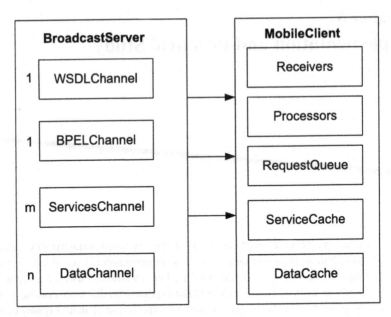

Fig. 8.1 Testbed - Component View

- One data index channel (only required by index channel method)

Each channel consists of a number of *buckets*. A bucket is the smallest logical unit of a broadcast channel. Broadcast contents are stored in buckets. The following procedure shows how the broadcast server works:

1. **Generate data repository**: A specified number of data sets are generated and stored in a data repository. These data sets are referenced by different services to provide information to end users.
2. **Construct data channels**: A specified number of data channels are created and all data sets from the data repository are placed in these channels.
3. **Construct the data index channel**: Each index entry in this channel contains a triplet (d, c, p), where k is the data set name, c is the channel number and p is the sequence number of the starting bucket for the data set. A bucket in this channel contains a number of index entries.
4. **Generate services repository**: A specified number of services are created and stored in a services repository. These are simple M-services to be delivered to mobile users through services channels. Each service references a data set. A specified distribution is used to determine which data set is referenced by a service.
5. **Construct services channels**: A specified number of services channels are created and all services from the services repository are placed in these channels.
6. **Construct the services index channel**: Each index entry in this channel contains a triplet (k, c, p), where k is the service key, c is the channel number and p is the

sequence number of the starting bucket for an M-service. A bucket in this channel contains a number of index entries.

7. **Construct the BPEL channel**: A specified number of composite services are created. The BPEL record for each composite service is generated based on the following user specified parameters:

- Composite service type
- Number of required services
- Minimum number of children per node
- Maximum number of children per node

There are three supported composite service types, namely *sequential, parallel,* and *hybrid*. A hybrid composite service may contain both sequential and parallel services. The number of required services determines how many services are referenced by a composite service. A random number is generated between the minimum and maximum number of children per node to determine the number of child nodes at each node in the BPEL tree. If the minimum number of children is greater than one, it is possible for the last non-leaf node to have less than minimum number of child nodes.

8. **Start broadcasting**: After all required broadcast channels are created, the broadcast sever start broadcasting the contents in each channel. When the end of a channel is reached, it will start from its beginning again.

8.1.2 Mobile Client

A mobile client is used to simulate a mobile device. It is responsible for accessing services from broadcast channels and record the time measurements. The following process explains how a mobile client works:

1. **Generate request**: A request is generated by selecting a composite service based on specified distribution.
2. **Obtain BPEL record**: Once a composite services is selected, its BPEL record is retrieved from the BPEL channel.
3. **Execute composite service**: The BPEL tree is traversed. At each node, the following process will occur:

- A service request is generated and placed in the request queue.
- The service request will be picked up when there is a free receiver.
- The receiver will try to retrieve the selected service from services channels based on the specified physical access method.
- Once retrieved, the service is placed in the services cache.
- The service will then be picked up by a free processor.
- The processor will execute the service and generate a data request in the request queue.
- The data request will also be picked up by a free receiver.

- The receiver will try to retrieved the requested data and place the retrieved data in the data cache. The retrieved data might be used by other services.

4. **Record time measurements**: At last all the time measures gathered during the execution process are written into an output file with a specific format.

8.2 Experiments

In this section, we present some experiments we conducted using the developed testbed. For each experiment, we first show the experiment settings. Then we present and analyze the results.

As already discussed, the total access time is the accumulated access times spent by all receivers. Since multiple receivers may work in parallel, the total access time does not reflect the total response time for a request to be fulfilled. For this purpose, we introduce a new factor, *transaction time*. The transaction time is defined as the total time elapsed from the point that a user has entered required parameters till the point that the required data is retrieved. The transaction time is an important factor for studying the behavior of different access methods and channel organizations under different semantics. It reflects the total time a user has to wait for a request to be fulfilled after a service is selected. First, let us look at simulation settings, as shown in Table 8.1, for all experiments presented in this section.

Table 8.1 Simulation settings for composite M-services

Number of services	1000 - 10000
Number of service channels	10
M-services size	1K - 5K bytes
Number of data channels	10
Data sets size	1K - 5K bytes
Broadcast bucket size	512 bytes
Data rate	20M bps
Number of total nodes per BPEL tree	10
Number of BPEL children per node	1 - 4
Number of data sets per service	5
Service key size	16 bytes
Offset size	8 bytes
Data request distribution	Zipf
Service request distribution	Zipf
Receiver pool size	1 - 10
Execution pool size	1 - 10
Services cache size	0 - 50 KB
Confidence level	0.95
Confidence accuracy	0.05

The number of nodes per BPEL tree in the table indicates the number of simple services required by a composite service. The number of BPEL children per node defines the number of child nodes any BPEL node could have. This setting is only used by hybrid composite services. When a hybrid composite service is generated, its BPEL structure is restricted by this setting. As already mentioned, recent wireless technologies such as Wi-Fi and WiMAX, have promised high data rates in the order of tens to hundreds of mega bps [14]. The data rate is expected to further increase in the near future. In these experiments, the data rate of a broadcast channel is assumed to be 20Mbps. The *Confidence level* and *Confidence accuracy* in the table are used to control the accuracy of the simulation results. Users can specify the values of confidence level and accuracy before starting simulation. The simulation will not end until the expected confidence level and accuracy are achieved. By using these two parameters, we ensure that simulation results we obtained are stable.

8.2.1 Comparison of Analytical and Simulation Results

In this section, we compare the analytical and simulation results of different channel organizations. By comparing the analytical results to simulation results, we are able to prove the correctness of the analytical models derived for different channel organizations. Table 8.2 shows the simulation settings for this experiment.

Table 8.2 Simulation settings for comparing analytical and simulation results

Number of services	10000 - 100000
Number of service channels	1
Number of data channels	1
Data rate	200M bps
Number of nodes per BPEL tree	10
Number of data sets per service	1
Composite service type	Sequential
Receiver pool size	1
Execution pool size	1
Services cache size	0 KB

As can be seen from the table, the experiment executes a sequential composite service consisting of 10 simple services using different channel organizations. Receiver pool size is set to 1 which means only one service or data set can be accessed at a time. With execution pool size also set to 1, only one service can be executed at a time. Services cache size is 0 KB which implies no service can be cached locally to be executed later. Experiment results are shown in Figures 8.2. In the figures, analytical results are marked as *(A)* and simulation results are marked as *(S)*.

Figure 8.2 shows the comparison for total service access times. As shown in the figure, the analytical results match quite well with the simulation results. Further-

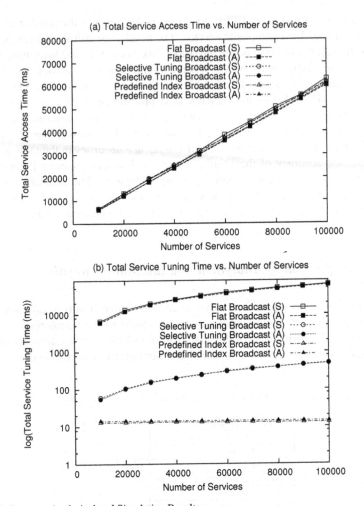

Fig. 8.2 Compare Analytical and Simulation Results

more, all three channel organizations yield similar access time performance. This is because the access time performance is only affected by the broadcast cycle and the number of requested services. The number of requested services is a constant in this experiment. The broadcast cycle for these three channel organizations are almost identical. The only difference is the extra offset values introduced by the selective tuning which is negligible in comparison to the size of a broadcast cycle. This explains why these three channel organizations demonstrate close service access time performance. Let us now look at the tuning time performance. The tuning times for flat broadcast are much larger than the other two methods. Logarithm is applied to all the tuning time values for the flat broadcast to fit in the same figure with the other two methods. Again, the analytical results match well with the simulation results. This proves the correctness of the analytical models. The figures also shows big differences in service tuning time performance for these three channel organizations. The difference is caused by how services are retrieved. For flat broadcast, a receiver must stay active all the time during the service retrievals. The total tuning time should be identical to the total access time. For selective tuning, the receiver only need to read the headers of services in order to find the requested services. This would significantly reduce the total tuning time compared to the flat broadcast. Lastly, for predefined index, the receiver already knows the locations of each service its asks for. Therefore, the receiver only needs to be active when it first tunes into the broadcast channel to determine its tune-in position and when it is downloading a service. The predefined index method should demonstrate the best tuning time performance. Its total tuning time is not affected by the total number of services. The experiment results shown in Figure 8.2 supports this analysis.

8.2.2 Comparison of Different Traversing Algorithms

Three different algorithms are proposed in this book for traversing access action trees, namely, *depth-first*, *breadth-first*, and *closest-first*. In this experiment, we compare these three algorithms. We also study the behaviour of different types of composite services with varying service cache size. We consider three type of composite services, *sequential*, *parallel* and *hybrid*. For a sequential composite service, all required services must be executed sequentially. For a parallel composite service, all required services could be executed concurrently. A hybrid service is the combination of both. Table 8.3 shows the simulation settings for this experiment.

Table 8.3 Simulation settings for comparing BPEL traversing algorithms

Number of services	5000
Composite service type	Sequential/Parallel/Hybrid
Receiver pool size	1
Execution pool size	1
Services cache size	0 - 30KB

Figure 8.3 shows the simulation results for this experiment. We make the following observations from the simulation results:

- With the same simulation settings, the sequential service type always demonstrates the worst transaction time. This is because all required simple services must be executed sequentially.
- Closest-first algorithm shows the best transaction time performance. This is because this algorithm always tries to retrieve the closest service or data whenever possible. As result, the overall transaction time is much shorter that the other two algorithms.

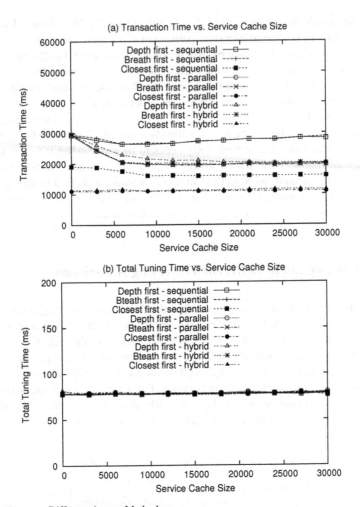

Fig. 8.3 Compare Different Access Methods

- Transaction time decreases with service cache size and stabilizes when the service cache size reaches certain point. This is because when service cache size increases, more services could be saved for later execution even if its dependent services are not executed yet. This certainly helps the transaction time performance. However, the performance only improves to a point that there is enough service cache to save those services that are not ready to be executed.
- The service cache size has less impact on the closest-first algorithm than the other two algorithms. This is because data items required by a service are retrieved faster with the closest-first algorithm. As a result, services go through the cache faster.

- All combinations of service types and traversing algorithms demonstrate similar total tuning time performance. The total tuning time is only decided by two factors, channel organization and number of required services and data items. Both factors are constants in this experiment. Therefore, all combinations show similar total tuning times.

8.2.3 Impact of Client Semantics

In this section, we present a set of experiments conducted for studying the impact of client semantics for different service types. We also study the impact of the following factors on the access efficiency:

- Number of services
- Number of required simple services per composite service
- Number of required data sets per simple service

For each experiment, we use two different client semantic settings as shown in Table 8.2.3.

Table 8.4 Simulation settings for comparing different semantics

Semantics-1		Semantics-2	
Receiver pool size	1	Receiver pool size	10
Execution pool size	1	Execution pool size	10
Services cache size	0 KB	Services cache size	500 KB

The *Semantics-1* settings would force all services to be accessed and executed sequentially. The *Semantics-2* settings would allow all services to be accessed and executed at the same time.

Number of services

In this experiment, we study the impact of total number of services on the access efficiency. Figure 8.4 shows the simulation results under different semantics. As can be seen from Figure 8.4(a), all transaction times increase linearly with the number of total services. This is because the broadcast cycle increases with the number of total services and it takes longer to retrieve a service when the broadcast cycle increases. The figure also shows that all three types demonstrate similar transaction time performances under Semantics-1. This is because even though hybrid and parallel services allow some or all services be retrieved and executed at the same time, there is not enough system resources for parallel retrieval or execution according to the semantics. Therefore, all services must be retrieved and executed in sequence. This makes hybrid and parallel services effectively the same as sequential services. Under Semantics-2, however, the sequential type has the worst transaction time performance and the parallel type has the best. This is because the semantics allow services to be retrieved and executed in parallel whenever possible. But for a sequential service, all services still have to be retrieved and executed sequentially regardless of the client semantics.

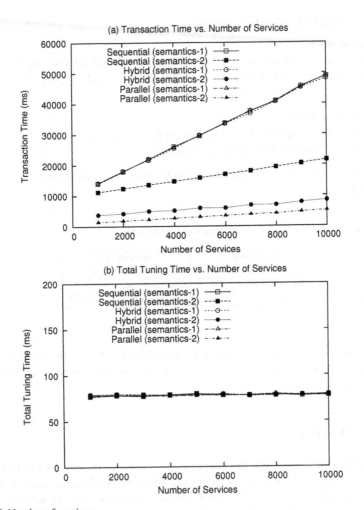

Fig. 8.4 Number of services

Figure 8.4(b) shows the comparison for the total tuning times. As shown in the figure, all different types demonstrate similar tuning time performances for both semantics. This is because the tuning time performance is only affected by two factors, the number of required simple services and the number of required data sets.

Both factors are constants in this experiment. Therefore, all composite service types demonstrate the same total tuning time performance regardless of the semantics.

Number of required simple services

In this experiment, we study the impact of number of required simple services per composite service. Figure 8.5 shows the simulation results for both semantics. According to Figure 8.5(a), all transaction times increase with the number of required simple services. It is intuitive that requesting for more services would lead to longer transaction time. Semantics-1 forces all simple services to be accessed sequentially. As a result, all service types show the same transaction time performance under Semantics-1. On the other hand, parallel services demonstrate the best transaction time performance under Semantics-2 because all required simple services can be accessed at the same time.

Figure 8.5(b) shows the same tuning time performance for both Semantics-1 and Semantics-2. This is because the total tuning time is the sum of the tuning times of all receivers. Adding more receivers does not decrease the total tuning time but instead spread the tuning time across all the receivers.

Number of required data sets

In this experiment, we study the impact of the number of required data sets per simple service. Simulation results are shown in Figure 8.6.

According to Figure 8.6(a), transaction times for Semantics-1 increase linearly and show the same performances for all service types. This is because it takes longer to retrieve more data sets sequentially. Under Semantics-2, though, transaction times hardly increase with the number of required data sets. This is because regardless of the service type the defined semantics would allow all data sets to be retrieved concurrently. Parallel services show the best transaction time performance because all their simple services could be accessed in parallel as well. All service types show the same tuning time performance in Figure 8.6(b) for both Semantics-1 and Semantics-2. This is because both the number of required services and data sets are the same for all service types.

8.2.4 Summary

Based on the experiments presented above, we note the following observations: (1) Different channel organizations have little impact on transaction time because the overhead introduced by extra index in broadcast channels is negligible compared to the whole broadcast cycle; (2) Channel organizations have huge impact on tuning time because they determine how long mobile devices would stay active during a

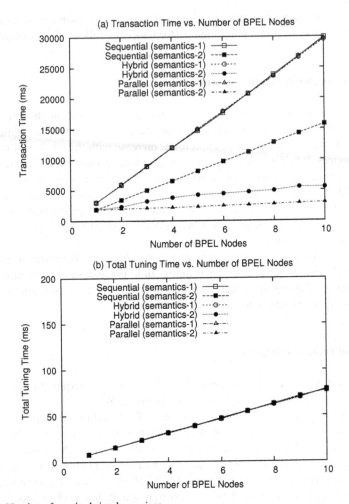

Fig. 8.5 Number of required simple services

transaction; (3) Flat broadcast always demonstrates the worst tuning time performance and index based methods always demonstrate the best tuning time performance; (4) Close-first algorithm has the best access performance because it takes
into consideration the locations of M-services; (5) In general, sequential composite
services show worse performance than parallel and hybrid composite services when
the number of required services and data sets are the same; (6) Special client semantics could force parallel composite services to execute in sequential mode and thus
have the same performance as sequential composite services; (7) Tuning time performance is not affected by service types but instead only by the number of required
services and data sets.

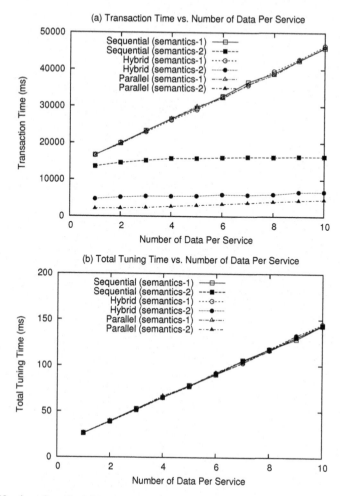

Fig. 8.6 Number of required data sets

Chapter 9
Open Problems

In this chapter, we discuss some key open problems for accessing M-services in wireless broadcast environments.

9.1 Complex Composite Services

In this book, we assume that supported composite services are pre-defined and can only contain simple services. This means all required services and their locations can be determined based on the BPEL and index information. In reality, however, composite services could be much more complex. For example, a composite service could invoke other composite services. This would result in a more complicated invocation sequence. Furthermore, only two types of BPEL activities, *sequence* and *flow*, are considered for composite services covered in this book. BPEL activities such as *if*, *while*, *forEach*, and *repeatUntil* ([68]) could produce more complex composite services than the supported types in this book. For example, an *if* activity indicates that some invoked services may not be determined until the invoking services have been executed. This would require access methods to update the access sequence based on which services are to be invoked. Some other BPEL activities such as *while* could require some services to be executed repeatedly. In such case, these services might need to be cached locally for a longer period of time to avoid frequent retrievals from broadcast channels.

9.2 Access Patterns

In an M-services system, some services could be accessed more frequently than the others. It is useful to investigate the impact of different access patterns. In this book, we assume service and data access follows Zipf distribution. However, we do not provide a comparison on access efficiency for different access patterns. Different

X. Yang and A. Bouguettaya, *Access to Mobile Services*, Advances in Database Systems 38, DOI: 10.1007/978-0-387-88755-5_9, © Springer Science + Business Media, LLC 2009

data access methods were proposed in the past to support skewed data access in data-oriented wireless broadcast systems [1, 2, 62, 75]. A typical approach to improve access efficiency in such systems is to broadcast frequently accessed data records more often than less popular ones. Channel organizations proposed in this book assume services and data sets are uniformly distributed in broadcast channels. It would be interesting to experiment with such organizations in service-centric broadcasts.

9.3 Service Composition Patterns

A composite service consists of one or more services. In this book, a composite service is created by randomly selecting a specified number of services and construct the BPEL definition based on the specified type, such as sequential, parallel or hybrid. In reality, some services might be used more often than others. It is important to study such service composition patterns and investigate their impact on access efficiency. Furthermore, similar to access pattern considerations, new channel organizations might be needed to achieve best access efficiency for different service composition patterns.

9.4 Automatic Service Composition

In this book, we assume composite services are pre-defined. The corresponding BPEL records for all composite services are made available on the BPEL channel. In recent years, a lot of research work has been conducted toward automatic service composition based on semantics in wired networks [37, 40, 4, 3, 39, 41]. Automatic service composition means composite services are not pre-defined but instead are created "on-the-fly" based on user input and service semantics. Automatic service composition requires dealing with three major research thrusts: semantic description of Web services, composability of participant services, and generation of composite service descriptions. To support automatic service composition for broadcast based M-services, semantic descriptions used for service compositions must also be made available on broadcast channels. Mobile devices need to access semantic descriptions first to generate required service compositions based on user requests. Then mobile devices would retrieve required services defined by the generated service compositions from broadcast channels and execute them.

References

1. S. Acharya, R. Alonso, M. Franklin, and S. Zdonik. Broadcast Disks: Data management for Asymmetric Communication Environments. In *Proceedings of the International ACM SIGMOD Conference on Management of Data*, pages 199–210, San Jose, CA, May 1995.
2. S. Acharya, M. Franklin, and S. Zdonik. Dissemination-based Data Delivery Using Broadcast Disks. *IEEE Personal Communications*, 2(6), 1995.
3. M. S. Akram, B. Medjahed, and A. Bouguettaya. Supporting Dynamic Changes in Web Service Environments. In *Proceedings of the International Conference on Service Oriented Computing*, pages 319–334, Trento, Italy, December 2003.
4. B. Benatallah, B. Medjahed, A. Bouguettaya, A. Elmagarmid, and J. Beard. Composing and Maintaining Web-based Virtual Enterprises. In *Proceedings of the International Workshop on Technologies for E-Services*, pages 155–174, Cairo, Egypt, September 2000.
5. T. Berners-Lee, R. Fielding, and L. Masinter. *RFC 2396: Uniform Resource Identifiers (URI): Generic Syntax*, August 1998.
6. R. Bort and G. R. Bielfeldt. *Handbook of EDI*. Warren, Gorham and Lamont, Boston, Massachusetts, 1994.
7. F. Casati and M.-C. Shan. Models and Languages for Describing and Discovering E-Services (Tutorial). In *Proceedings of the International ACM SIGMOD Conference on Management of Data*, May 2001.
8. Y. Chehadeh, A. Hurson, and L. Miller. Energy-Efficient Indexing on a Broadcast Channel in a Mobile Database Access System. In *International Symposium on Information Technology (ITCC)*, Las Vegas, NV, USA, March 2000.
9. M.-S. Chen, P. S. Yu, and K.-L. Wu. Indexed Sequential Data Broadcasting in Wireless Mobile Computing. In *International Conference on Distributed Computing Systems (ICDCS)*, USA, May 1997.
10. T. H. Cormen, C. E. Leiserson, R. L. Rivest, and C. Stein. *Introduction to Algorithms*, chapter 22.3: Depth-first search, pages 540–549. MIT Press and McGraw-Hill, second edition, 2001.
11. T. H. Cormen, C. E. Leiserson, R. L. Rivest, and C. Stein. *Introduction to Algorithms*, chapter 22.2: Breadth-first search, pages 531–540. MIT Press and McGraw-Hill, second edition, 2001.
12. F. Curbera, M. Duftler, R. Khalaf, W. Nagy, N. Mukhi, and S. Weerawarana. Unraveling the Web Services Web: An Introduction to SOAP, WSDL, and UDDI. *IEEE Internet Computing*, 6(2), February 2002.
13. R. Fielding, J. Gettys, J. Mogul, H. Frystyk, and T. Berners-Lee. *RFC 2068: Hypertext Transfer Protocol - HTTP/1.1*, January 1997.
14. N. Fourty, T. Val, P. Fraisse, and J.-J. Mercier. Comparative analysis of new high data rate wireless communication technologies "From Wi-Fi to WiMAX". In *Joint International Conference on Autonomic and Autonomous Systems / International Conference on Networking and Services (ICAS/ICNS)*, page 66, Papeete, Tahiti, 2005.

15. D. J. Goodman. The Wireless Internet: Promises and Challenges. *IEEE Computer*, 33(7), July 2000.

16. S. Hambrusch, C.-M. Liu, W. G. Aref, and S. Prabhakar. Query Processing in Broadcasted Spatial Index Trees. *Lecture Notes in Computer Science*, 2121:502, 2001.

17. Q. Hu, D. Lee, and W. Lee. A comparison of indexing methods for data broadcast on the air. In *Proceedings of the 12th International Conference on Information Networking (ICOIN–12)*, pages 656–659, January 1998.

18. Q. Hu, D. L. Lee, and W.-C. Lee. Indexing Techniques for Wireless Data Broadcast under Data Clustering and Scheduling. In *ACM Conference on Information and Knowledge Management (CIKM)*, Kansas City, Missouri, USA, November 1999.

19. Q. Hu, D. L. Lee, and W.-C. Lee. Power Conservative Multi-Attribute Queries on Data Broadcast. In *International Conference on Data Engineering (ICDE)*, pages 157–166, San Diego, California, USA, February/March 2000.

20. Q. Hu, W.-C. Lee, and D. L. Lee. A Hybrid Index Technique for Power Efficient Data Broadcast. *Journal on Distributed and Parallel Databases*, 9(3):151–177, March 2001.

21. Q. Hu, W.-C. Lee, and D. L. Lee. Indexing Techniques for Power Management in Multi-Attribute Data Broadcast. *Mobile Networks and Applications (MONET)*, 6(2):185–197, 2001.

22. Y. Huang and Y. H. Lee. An Efficient Indexing Method for Wireless Data Broadcast with Dynamic Updates. In *Proceedings of the 1st IEEE ICCCAS Conference*, Chengdu, China, June 2002.

23. T. Imielinski and H. F. Korth, editors. *Mobile Computing*. Kluwer Academic Publishers, 1996.

24. T. Imielinski, S. Viswanathan, and B. R. Badrinath. Energy Efficient Indexing on Air. In *Proceedings of the International ACM SIGMOD Conference on Management of Data*, pages 25–36, 1994.

25. T. Imielinski, S. Viswanathan, and B. R. Badrinath. Power Efficient Filtering of Data an Air. In *International Conference on Extending Database Technology (EDBT)*, pages 245–258, UK, March 28–31 1994.

26. T. Imielinski, S. Viswanathan, and B. R. Badrinath. Data on Air: Organization and Access. *IEEE Transactions on Knowledge and Data Engineering (TKDE)*, 9(3):353–372, 1997.

27. R. Kalakota and A. B. Whinston. *Frontiers of Electronic Commerce*. Addison Wesley (ISBN: 0-201-84520-2), February 2000.

28. G. Kollios, D. Gunopulos, and V. J. Tsotras. On Indexing Mobile Objects. In *Proceedings of the Eighteenth ACM SIGACT-SIGMOD-SIGART Symposium on Principles of Database Systems*, pages 261–272, Philadelphia, Pennsylvania, May 31–June 2 1999.

29. W.-C. Lee and D. L. Lee. Using Signature Techniques for Information Filtering in Wireless and Mobile Environments. *Special Issue on Databases and Mobile Computing, Journal on Distributed and Parallel Databases*, 4(3):205–227, 1996.

30. Frank Leymann and Dieter Roller. Modeling business processes with bpel4ws. In *Modellierung*, 2004.

31. M. C. Little and D. L. McCue. Construction and Use of a Simulation Package in C++. Technical Report 437, Computing Science Technical Report, University of Newcastle upon Tyne, July 1993.

32. S.-C. Lo and A. L. P. Chen. An Adaptive Access Method for Broadcast Data under an Error-Prone Mobile Environment. *IEEE Transactions on Knowledge and Data Engineering (TKDE)*, 12(4):609–620, July/August 2000.

33. Z. Maamar, B. Benatallah, and Q. Z. Sheng. Towards a Composition Framework for E-/M-Services. In *ACM Workshop on Ubiquitous Agents on Embedded, Wearable, and Mobile Devices*, July 2002.

34. Z. Maamar, W. Mansoor, and Q. Mahmoud. Software Agents to Support Mobile Services. In *International Conference on Autonomous Agents and Multi Agents Systems*, July 2002.

35. Z. Maamar, W. Mansoor, and H. Yahyaoui. E-Commerce through Wireless Devices. In *IEEE International Workshops on Enabling Technologies: Infrastructure for Collaborative Enterprises*, June 2001.

36. R. Malladi and D. P. Agrawal. Current and future applications of mobile and wireless networks. *Communications of the ACM*, 45(10):144–146, October 2002.

37. B. Medjahed, B. Benatallah, A. Bouguettaya, and A. Elmagarmid. WebBIS: A Framework for Agile Integration of Web Services. *International Journal of Cooperative Information Systems*, 13(2), June 2004.

38. B. Medjahed, B. Benatallah, A. Bouguettaya, A. H. H. Ngu, and A. K. Elmagarmid. Business-to-business interactions: issues and enabling technologies. *The VLDB Journal*, 12(1):59–85, 2003.

39. B. Medjahed, A. Bouguettaya, and A. Elmagarmid. Composing Web Services on the Semantic Web. *The VLDB Journal*, 12(4):333–351, November 2003.

40. B. Medjahed, A. Bouguettaya, and M. Ouzzani. Semantic Web Enabled E-Government Services. In *Proceedings of the NSF Conference for Digital Government Research*, pages 250–253, Boston, Massachussets, USA, May 2003.

41. B. Medjahed, M. Ouzzani, and A. Bouguettaya. Using Web Services in E-Government Applications. In *Proceedings of the NSF Conference for Digital Government Research*, pages 371–376, Los Angeles, California, USA, May 2002.

42. Object Management Group (OMG). *The OMG's CORBA Website*. http://www.corba.org/.

43. G. D. Parrington, S. K. Shrivastava, S. M. Wheater, and M. C. Little. The Design and Implementation of Arjuna. *Computing Systems*, 8(2):255–308, 1995.

44. T. Pilioura, A. Tsalgatidou, and S. Hadjiefthymiades. Scenarios of Using Web Services in M-Commerce. *ACM SIGecom Exchanges*, 3(4), January 2003.

45. J. Postel. *RFC 318: Telnet protocol*, April 1972.

46. J. Postel and J. K. Reynolds. *RFC 959: File transfer protocol*, October 1985.

47. Forrester Research. *IT Facts*. http://www.itfacts.biz, 2005.

48. S. Saltenis and C. S. Jensen. Indexing of Moving Objects for Location-Based Services. In *International Conference on Data Engineering (ICDE)*, pages 463–472, 2002.

49. J. A. Senn. The Emergence of M-Commerce. *IEEE Computer*, 33(12), December 2000.

50. N. Shivakumar and S. Venkatasubramanian. Efficient Indexing for Broadcast Based Wireless Systems. *Mobile Networks and Applications (MONET)*, 1(4):433–446, 1996.

51. Sun Microsystems. *Enterprise JavaBeans Technology (EJB)*. http://java.sun.com/products/ejb/.

52. Sun Microsystems. *Java Remote Method Invocation (Java RMI)*. http://java.sun.com/products/jdk/rmi/.

53. K.-L. Tan and J. X. Yu. Energy Efficient Filtering of Nonuniform Broadcast. In *International Conference on Distributed Computing Systems (ICDCS)*, 1996.

54. K. Tanaka, S. Ghandeharizadeh, and Y. Kambayashi, editors. *Information Organization and Databases*, chapter Power Conserving and Access Efficient Indexes for Wireless Computing. Kluwer Academic Publishers, 2000.

55. J. Tayeb, O. Ulusoy, and O. Wolfson. A Quadtree-Based Dynamic Attribute Indexing Method. *The Computer Journal*, 41(3):185–200, 1998.

56. The World Wide Web Consortium (W3C). *Extensible Markup Language (XML)*. http://www.w3.org/XML/.

57. The World Wide Web Consortium (W3C). *Simple Object Access Protocol (SOAP)*. http://www.w3.org/TR/soap.

58. The World Wide Web Consortium (W3C). *Web Services Architecture*. http://www.w3.org/TR/ws-arch/.

59. The World Wide Web Consortium (W3C). *Web Services Description Language (WSDL)*. http://www.w3.org/TR/wsdl.

60. The World Wide Web Consortium (W3C). *Web Services Description Requirements* *http://www.w3.org/TR/ws-desc-reqs/*, October 2002.

61. UDDI.org. *Universal Description, Discovery, and Integration (UDDI)*. http://www.uddi.org.

62. N. Vaidya and S. Hameed. Scheduling data broadcast in asymmetric communication environments. *ACM Baltzer Wireless Networks*, 5(3):171–182, 1999.

63. U. Varshney and R. J. Vetter. Mobile Commerce: Framework, Applications and Networking Support. *Mobile Networks and Applications (MONET)*, 7(3), June 2002.

64. U. Varshney and R. J. Vetter, editors. *Special Issue on Mobile Commerce*, Mobile Networks and Applications (MONET) 7(3), June 2002.

65. U. Varshney, R. J. Vetter, and R. Kalakota. Mobile Commerce: a New Frontier. *IEEE Computer*, 33(10):32–38, October 2000.

66. G. F. Welch. A Survey of Power Management Techniques in Mobile Computing Operating Systems. *Operating Systems Review*, 29(4):47–56, 1995.

67. J. E. White. *RFC 524: Proposed mail protocol*, June 1973.

68. WSBPEL2.0.   *Web Services Business Process Execution Language Version 2.0.* http://docs.oasis-open.org/wsbpel/2.0/OS/wsbpel-v2.0-OS.html.

69. X. Yang and A. Bouguettaya. Broadcast-Based Data Access in Wireless Environments. In *International Conference on Extending Database Technology (EDBT)*, pages 553–571, 2002.

70. X. Yang and A. Bouguettaya. Adaptive Data Access in Broadcast-Based Wireless Environments. *IEEE Transactions on Knowledge and Data Engineering (TKDE)*, 17(3):326–338, 2005.

71. X. Yang and A. Bouguettaya. Using a Hybrid Method for Accessing Broadcast Data. In *IEEE International Conference on Mobile Data Management (MDM)*, pages 38–47, New York, NY, USA, 2005. ACM Press.

72. X. Yang and A. Bouguettaya. Efficient Access to Wireless Web Services. In *IEEE International Conference on Mobile Data Management (MDM)*, page 30, 2006.

73. X. Yang and A. Bouguettaya. Semantic Access to Multi-Channel M-services. *IEEE Transactions on Knowledge and Data Engineering (Accepted)*, 2008.

74. X. Yang, A. Bouguettaya, B. Medjahed, H. Long, and W. He. Organizing and Accessing Web Services on Air. *IEEE Transactions on Systems, Man, and Cybernetics, Part A (SMC-A)*, 33(6):742–757, 2003.

75. J. X. Yu and K. L. Tan. An Analysis of Selective Tuning Schemes for Non-uniform Broadcast. *Data and Knowledge Engineering*, 22(3):319–344, 1997.

Index